The Kettle Begins To Simmer 2012

The Imbalanced Atmosphere Chronicles

A Connective Collaboration of Fiction

by

Deborah Lapping Dorsey

and

William Dorsey

The Kettle Begins to Simmer 2012

All Rights Reserved.

Dedication

We dedicate this book to those members of humanity, young and old, male and female, who have committed their lives to preserving and protecting our fragile planet and the living organisms that call it home.

Introduction

Dear Reader:

As you embark upon the reading of this novel, be prepared to discover structural differences that will set it apart from most of the other novels that you may have read. In searching through my books of literary terms, we could find nothing that described what my husband William and I have put together. Consequently, we have coined the term *A Connective Collaboration of Fiction* to delineate our work.

A Connective Collaboration of Fiction is a work in which two (or more) authors are writing separate narratives about the same events and some of the same characters intertwined in the same time frame. All of the characters may not know all of the other characters, but they all are connected to one another in some fashion. In this novel, *setting* is the unifying factor underlying the major conflict that each group of characters must face.

There is not one central character, or protagonist, in this book, but rather a series of characters to whom you will be introduced. Some of the characters you will hear from several times, while others you will meet only briefly for now, but will encounter again in the next of our series, <u>Fear In The Dust</u>.

Although this is fiction, authenticity is an important element in our writing, particularly as it pertains to the facts, figures, and quotes used in the reports submitted by the character Scott Wasson. While Scott is fictitious, his reports are essentially factual. Mrs. Starkey, another fictitious character, also gives a brief history of the leadership of the Environmental Protection Agency since its inception. That information can be found on various Internet websites, including in the Environmental Protection Agency Archives.

The other term we wish to introduce to you is *Imbalanced Atmosphere* or *Atmospheric Imbalance*. Again, this is a term we have coined to describe the state of our planet today. Many years ago, when people first began to notice that the composition of the atmosphere surrounding the Earth was changing, there were countless studies and discussions about the dangers of Greenhouse Gases. Then connections were made between Greenhouse Gases and the possibility of Global Warming. From Global Warming came the frightening issue of Climate Change and its current and potential adverse impact on our lives, from the grocery store to the sea shore. In the midst of all this, the critical discussions pertaining to the health of our fragile planet became politicized.

It is our hope that the term *Imbalanced Atmosphere* or *Atmospheric Imbalance* is sufficiently descriptive yet politically neutral enough that all facets of the political spectrum would be comfortable with its use. We all need to see action taken to regain the Earth's natural balance because, after all, this is our home. In the vastness of the universe, it is the only lifeboat we have, and we're all in it!

William Dorsey

Deborah Lapping Dorsey

Statesville, North Carolina

November 11, 2012

www.imbalanced-atmosphere-chronicles.com

Table of Contents

Locations and Characters

Chapter One
North Carolina 2002
"Leap of Faith"

Chapter Two
Environmental Protection Agency,
Washington, D. C.
Friday, May 25, 2012, 11:06 AM

Chapter Three
New Orleans, Louisiana 2005
"Jordan's Escape"

Chapter Four
Logan's Roadhouse, Fairfax, Virginia
Saturday, May 26, 2012, 8:44 PM

Chapter Five
West Plains, Missouri 2012
"Harvey and Grace"

Chapter Six
Environmental Protection Agency,
Washington, D. C.
Friday, June 15, 2012, 10:03 AM

Chapter Seven

Arlington, Texas 2012

"Nick's Dilemma"

Chapter Eight

Video Conference, Joplin, MO -

Washington, D. C.

Monday, June 18, 2012, 9:02 AM

Chapter Nine

Joplin, Missouri 2011

"Jordan's Near Miss"

Chapter Ten

Video Conference, Joplin, MO -

Washington, D. C.

Monday, July 2, 2012, 8:57 AM

"State of the Heartland: Joplin, Missouri"

Chapter Eleven

South-Central Nebraska 2012

"Miller's Trucking"

Chapter Twelve

Video Conference, Miner, MO -

Washington, D. C.

Monday, July 5, 2012, 9:08 AM

Chapter Thirteen

Joplin, Missouri 2012

"The Phone Call"

Chapter Fourteen

Video Conference, New Orleans -

Washington, D. C.

Monday, July 23, 2012, 9:01 AM

"State of the Heartland:

The Mississippi River Valley

and New Orleans, Louisiana"

Chapter Fifteen

Wichita, Kansas 2012

"What's Wrong With Betsy?"

Chapter Sixteen

Video Conference, Dallas, TX -

Washington, D. C.

Monday, August 6, 2012, 8:58 AM

"State of the Heartland: The West Nile Virus"

Chapter Seventeen

Arlington, Texas 2012

"The Parents Visit"

Chapter Eighteen
Video Conference, Hastings, NE -
Washington, D. C.
 Monday, August 20, 2012, 9:00 AM
"State of the Heartland: Heat and Drought"

Chapter Nineteen
Wichita, Kansas 2012
"What About the Farm?"

Chapter Twenty
South-Central Nebraska 2012
"The Playhouse"

Chapter Twenty-One
South-Central Nebraska 2012
"The Unexpected Conversation"

Chapter Twenty-Two
Environmental Protection Agency,
Washington, D. C.
Saturday, September 8, 2012, 2:03 PM

Chapter Twenty-Three
North Carolina 2012
"The Fire Pit"

Locations and Characters

Statesville, North Carolina

Vanessa Carson – retired teacher; C.B.'s wife; Nick's mother

C.B. Carson – retired; Vanessa's husband; Jordan's father; Bob's childhood friend; Harvey's younger brother; Katy Carson's great-uncle

Environmental Protection Agency, Washington D. C.

Alda Martin Starkey – Deputy Director, E. P. A. Air, Climate and Energy Research Program; wife of Admiral Starkey

Carter Jefferson Penley – Alda's Administrative Assistant

Patty McCarty – Alda's Executive Secretary; John McCarty's wife

Justin Moore – Alda's Research Resource

Scott Wasson – internationally renowned environmental scientist; deeply committed to Katy Carson

Katy (Katherine) Carson – Purdue graduate student; deeply committed to Scott Wasson

John McCarty – Patty McCarty's husband

Kaitlyn McCarty – John and Patty McCarty's teen-aged daughter

Rear Admiral William Henderson Starkey –
Alda's husband

New Orleans, Louisiana

Jordan Carson – C.B.'s daughter; Katy
Carson's cousin

Brent – Jordan's husband (briefly);
prescription drug dealer

Arlington/Dallas, Texas

Nick Patton – Environmental Specialist for the
state of Texas; Kara's husband; Ashley's
father; Vanessa's son

Kara Patton – Student; Nick's wife; Ashley's
mother

Ashley – Nick and Kara's daughter; Vanessa
and C.B.'s granddaughter; two years old

The Metroplex – The general area spanning
the entire Fort Worth/Dallas region

Fort Worth, Texas

Peggy – Vanessa's sister; Carol's mother

Carol – Vanessa's niece; Peggy's daughter;
dies as the result of a car accident

Joplin, Missouri

Jordan Carson Beeman – Student; Kyle's new
wife; C.B.'s daughter; Harvey's niece; Katy
Carson's cousin

Kyle Beeman – Computer whiz; Jordan's new husband

West Plains, Missouri

Harvey Carson – Elderly; Grace's husband; Katy Carson's grandfather; C.B.'s older brother; Jordan's uncle

Grace Carson – Retired; Harvey's second wife

Wichita, Kansas

Mike Wasson – Private airplane electrician and mechanic; Stacy's husband; Betsy's father; Scott Wasson's distant cousin

Stacy Wasson – Hair Stylist and salon owner; Mike's wife; Betsy's mother

Betsy – Mike and Stacy's eight-year-old daughter; contracts West Nile virus

South-Central Nebraska

Bob Miller – Owner, Miller's Trucking; Laura's husband; Jake's father; C.B.'s childhood friend

Laura Miller – Bob's wife; Jake's mother

Jake Miller – Bob and Laura's son; Mandy's husband

Mandy Miller – Jake's wife; from North Carolina; care-taker of all of the children

Alana Weatherford – Retired registered nurse; widow; helps in caring for the children

Hannah Ingersol – Clinical dietitian; hired to prepare meals for the children

Chapter One

North Carolina 2002

"Leap of Faith"

Some families live close enough to each other that life events are easily shared, and traditions and memories are easily made and passed down through the generations. For other families, life's circumstances scatter its members like autumn leaves in a strong breeze. With effort, those families, too, can remain united if they value that sense of togetherness which binds families, and they are willing to promote and support that oneness of spirit. They can also share their life events and create their own traditions and memories. In some ways, those moments are more precious because they are sometimes rare or difficult to achieve.

Then there are those families who fail to fit into either of the previous categories. Those are the ones who choose not to be close, physically or emotionally. Fortunately for Vanessa and C.B., neither of them fell into the third category. As much as they loved each other, they missed their families. Like so many others, they found themselves at retirement or close to it with their families scattered, and they knew that they were the ones who must bring about change if any of their family members were to be in their lives once more. They were again ready for roots, they wanted a home they could make their own, and they wanted family nearby. When they looked closely, they realized that they could never envision that future for themselves in Austin, Texas.

The answer seemed to unfold before them after Vanessa had returned from a visit to her mother and stepfather in South Carolina.

"I wish it weren't such an arduous trip to visit them," she complained in frustration as she placed her suitcase on the bed and began to unpack. "And the expense... arghhh... but it's not just the money. Their few remaining years left on this planet are just flying by, and I'm half-way across the country. I hate it! I just hate it!"

C.B. was sitting on the bed watching and listening and feeling a bit helpless. Suddenly, that playful twinkle in his eye and sheepish grin that always made Vanessa smile accompanied a clearing of his throat.

"Umm-hmm. How would you like to consider a move to the Carolinas? There's nothing holding us here, and I think spending our time with Caroline and Hugh is much more important than spending it with the rattlesnakes and scorpions."

Vanessa didn't know whether to laugh or cry, but she did know that what he was

2

saying was true and was exactly the right thing to do. All the forms, the paperwork, the Praxis, the references, the letters, the interviews... there would be so many details needing attention later. All she could think about at the moment was, "Yes! Yes! Let's do it!"

Vanessa had received major state and national recognition for excellence in teaching, and her credentials were impeccable, so she was not unduly concerned about finding employment. Still, it was a leap of faith to resign and head for the mountains without even a suggestion of a teaching position lined up. It was 2002, however, before the Great Recession, and such actions were not as suicidal as they might have proven to be just a few short years later.

They chose Statesville, some miles north of Charlotte, for myriad reasons. It was a smaller town with good-hearted people whose families had lived there for generations. In 1789 it had been incorporated as a village, and some of the tombstones that could still be read attested to inhabitants in the village much earlier than that. C.B. and Vanessa had often discussed with admiration the difficulties those early settlers must have endured as they carved a life out of the North American wilderness, and the couple honored the history of the area

3

and the perseverance and dedication to life that it represented.

As Vanessa was unpacking in North Carolina, she came across a book of her family's history compiled by her grandfather and great-aunt. In it she discovered the name of one of her grandmothers from some generations back, *Caroline M. Wood Bates,* who was born in North Carolina on January 17, 1825. Her mother had, in fact, been given this family name. No other information was given, so it would become her challenge to discover the actual birthplace of one of her ancestors. She was a native North Carolinian after all. How curious that after her many travels and travails, life would return her full circle to the home of one who appeared in her family line from so many generations past.

Statesville rested in the foothills of the Appalachian Mountains - close enough to have deeply rolling hills, forests of trees, and four distinct seasons, something they rarely experienced in Austin and missed from their own childhoods in the Midwest. The Blue Ridge Parkway was less than an hour away, and Love Valley was hidden just up the road, where people rode their horses to the feed store and rodeo and parked their pistols with the bartender at the local saloon.

Down the road past wineries, upscale Lake Norman, and NASCAR, sprawled

Charlotte - sparkling, fresh, new and old at the same time. It had money and class and style and an old-fashioned Southern charm that made visitors want to come back, and those who lived there want to stay. Best of all, Vanessa's parents lived just on the other side, in a small town that Hugh and his relatives from generations past had been instrumental in building. Vanessa could make the drive in two hours and did so many weekends.

They found a comfortable Southern Colonial ranch that was just right for them. Situated on almost four rolling acres of woods, it had everything they had been looking for: a circular drive, a stream, and a sense of history. It also had much work to be done, as it had been unoccupied for two years after its elderly owner had been moved to a nursing home. C.B. and Vanessa were undaunted and lovingly took on the task of restoring the home and grounds to their original beauty. Walking hand-in-hand across the meadow, sitting by the fire pit in the glen, or carving out a new trail on the ridge, C.B. and Vanessa never regretted their move or taking on such an all-consuming project.

By 2009 they had lost Vanessa's stepfather to cancer, her mother had suffered a series of small strokes and falls, some requiring stitches or physical therapy, the economy had collapsed, and Vanessa had been forced to retire from teaching as a result of increasingly painful back issues. *Had we not had the courage to move when we did,* she would think from time to time, *look at all we would have missed!*

§§§

Chapter Two

Environmental Protection Agency, Washington, D. C.
Friday, May 25, 2012, 11:06 AM

Alda Martin Starkey pensively eyed the red folder on the far left corner of the over-sized desk for the third time in as many minutes. Carter, her Administrative Assistant, had laid it gingerly in the designated hot spot ten minutes before, caught her eye as she finished her video conference, and directed her glance to the report under his hand. She had nodded acknowledgment, and, as he turned to leave, noticed the tinge of apprehension in his expression.

Well, the Deputy Director thought as she opened the file, *I'm ready for this. We've heard more than enough about this doctoral candidate boy wonder and his work.* Ten minutes later, Alda rose with a frown and moved to the windows. She stood for a long moment looking east toward the Hill, then walked to the open door of her office.

"Patty, I'll be unavailable for a bit, barring, of course, catastrophes of the first magnitude. And, please, have some tea brought up. Justin and Carter, please step in, and close the door behind you."

Alda Starkey's inner office staff exchanged looks and knew an issue of imminence was at hand. Her office door was rarely closed, as her staff was distanced well out of earshot except for the loudest of exchanges. Patty, Justin, and Carter shared an intense admiration and loyalty toward her, forged from long watching of the Deputy Director for the Research Program in action. They knew no interruptions, tea, and a closed door meant serious business indeed.

Alda, Justin, and Carter gathered at the smaller, round conversational table. As they settled in the comfortably-cushioned chairs, the Deputy Director started the discussion.

"The Scott Wakefield Wasson issue is now center stage on my horizon. Thank you, Carter, for the informative report. So, let's start with you. Based on all the input you've received, please give me your evaluation, concisely, of the reality with which we are dealing here."

Carter Jefferson Penley leaned forward and responded formally, elbows on his thighs with hands finger-tipped together, touching at chin level.

"Ma'am, Justin and I put ears-to-ground everywhere. Between his results and my reports, the background comments circulating from those knowledgeable appear to confirm, unanimously, the correctness of Scott Wasson's model and the accuracy of his projections."

The Deputy Director was silent, then pushed her chair back, and turned to gaze again across the distances to the Hill. Her spell was broken by the knock at the door, and Patty's immediate entrance with the covered tray taken from the hands of the delivering attendant. After serving the tea, Patty made inquisitive eye contact with each of the three, and, receiving no responsive reply, left the office, closing the door behind her.

Alda's attention returned to the conversation, and she turned toward Justin. "And you, man of magic, what do you bring to the table?"

Justin's reply was, as usual, informal and succinct. "Supportive of, and in agreement with Carter's assessment, plus two items, actually three, of interest. First, an unofficial copy of doctoral candidate Wasson's model and projections, for full analysis. An unnecessary effort, as it turns out, because, secondly, in face-to-face contact, Scott Wasson has offered all his research to us. Meaning, specifically, this office. And thirdly, he has requested a private audience with Air, Climate, and Energy Research

Program Deputy Director Alda Martin Starkey."

"Hmm... very good. Well done, Justin." After a moment's reflection, she asked pointedly, "Why us at this A.C.E. office? Why even the E.P.A.? Both of you have indicated Scott Wasson's work has generated considerable buzz throughout the environmental community, governmental and non-governmental organizations alike. You've talked with him. Tell me why you think Mr. Wasson has made this generous proposal. Also, give me a brief background of this young prodigy's credentials, as well as your impression of his character and personality."

"On his offer, best guess, several reasons," the Research Resource replied. "Wasson is extremely disappointed with the current inaccuracies in the Agency's preliminary September 2012 Report On The Environment, the automatic credibility it will receive when finalized and presented, and the action plans that will be put into place by many organizations world-wide based on these R.O.E. projections. He has recognized the embarrassment we would endure if the report would be discredited from outside sources, and that a timely revision from within the Agency would be the optimum solution for all concerned. And, finally, your reputation precedes you. He respects and trusts you."

Justin Moore shifted position slightly, and referred to the notes in the folder before him.

"Scott Wakefield Wasson, as you've said, is acknowledged to be extremely gifted. B.S. from M.I.T. at seventeen, focus in Mathematical Modeling and Numerical Analysis, completed in less than two years. Multi-disciplined environmental studies at three leading educational institutions with unequaled records of achievement.

"Currently a candidate in the Environmental Change Institute at Oxford for the Doctor of Philosophy in Geography and the Environment, an advanced research degree awarded to candidates who've already completed a significant piece of original research. Their doctoral program enables candidates to conduct further independent research, and he has completed three years in advanced research and applied development with several non-governmental organizations, including the Hadley Centre. Credited with visionary approaches in the field of atmospheric imbalance assessments, plus extraordinary innovations he has individually developed and applied. Widely regarded as among the very best. In some quarters, he is thought to be without equal.

"My one short meeting with him revealed several personality traits. First of all, Wasson is an outgoing type, engaging and open, an interesting conversationalist. Unquestionably bright and perceptive, prefers direct interaction, listens attentively before formulating replies, exhibits occasional irreverence to the extent of being outright brassy. He projects idealism, honesty, and generosity - professionally, at least - and an almost fearfully-based impatience. Every report indicates enviously universal admiration and respect. I haven't had time to delve into his more personal, umm, characteristics."

Alda nodded toward Justin. "Your investigations, for the moment, are sufficient. "But," with a nod toward her Administrative Assistant, "if this moves in a mutually-agreeable direction, as happened with Carter, I'll want a thorough knowledge of his background. The same type of interviews with family, friends and community, from the first years of schooling to the last."

She gave Carter full eye contact. "Before this Friday afternoon is done, set up a very fast-track process for Mr. Wasson to be able to join the Agency, should that possibility present itself. Entry at near the maximum we can offer a non-doctorate. GS fifteen, step five, with step ten after favorable evaluations at six months. Let me know if I need to talk to someone to move it along."

Then, to Justin, "See if Mr. Wasson would be available for a low-profile meeting tomorrow evening. A not-too-upscale restaurant, and not in D. C. Comfortable and informal, say a Ruby Tuesday, or better yet, someplace with a little ambient music. There's a Logan's Roadhouse about a half-hour out in Fairfax. Set it up for eight-thirty or nine o'clock, after most dining is done, but still with time for us to have a conversation. And please come along. To all appearances, this will be a late Saturday dinner with, perhaps, a mother, a son, and a friend."

§§§

Chapter Three

New Orleans 2005
"Jordan's Escape"

Jordan occasionally found herself, as have others along the way, with the wrong person in the wrong place at the wrong time. She had married Brent because the comfort and security of marriage and children had always held such an intense allure in her otherwise seemingly aimless life. *Besides, Jordan would think wistfully, Brent is nice to me, and I can live with that.* At almost thirty, she was keenly sensitive to her biological clock, and with no other prospects in sight, marriage to Brent did not seem to be such a bad idea.

Unfortunately, being an intelligent young woman did not protect her from being led down a dark, dangerous path by a devious man whose motives regarding her were malevolent and sinister. For him, a wife was superfluous; what he required was someone to run his scams with doctors in an unending effort to supply prescription drugs to an insatiable underground market. Thus, he began to entwine her in his seedy world, and by involving her, it seemed less likely that she would leave. How could she? She would be as culpable as he.

What Brent failed to recognize was Jordan's tenacious spirit. Dealing in drugs had never been her world. This was not what she had envisioned in her marriage, and it most certainly would not lead her to that cozy cottage with the white picket fence and a swing set in back. In addition, she was determined that she was not about to go to jail for someone who thought that the way to a better life was running the grungy streets of New Orleans dealing with the seamy underbelly of existence.

She squared her shoulders and resisted in every way she could. Her mother had recently died of cancer, and C.B., her dad and stepmother Vanessa, lived in North Carolina, so there was no family close enough to observe her dilemma. Besides, her pride made her reticent to reveal the magnitude of this fiasco to them. She was fighting the battle alone.

C.B. and Vanessa were sitting on the back steps of their North Carolina home when they placed the first of several calls to Jordan in New Orleans on Saturday morning, August 27, 2005.

"Jordan, you have to leave town," C.B. argued sternly but with a hint of pleading in his voice.

"Daddy, I'll be fine. People ride out storms around here all the time," she reassured him.

"But, honey, this is not just another storm. Katrina is a category five hurricane, and it's heading straight for you."

After several more equally fruitless calls, Vanessa and C.B. sat and shook their heads in numbed disbelief. How could she be so... *fill in the blank*... foolish, dumb, oblivious to the danger, stubborn, crazy... What was going on? Once they moved past their own panic and shock, they recognized that they must be the ones to take action in order to save their daughter, because, obviously, she was not going to save herself.

C.B. immediately packed their pick-up truck with everything he thought he might need for his rescue mission. He even made arrangements to obtain a U-Haul in New Orleans to help move her belongings, although he knew there would be precious little time at best for packing.

Tensely, Vanessa packed a cooler with food and drinks. It was normally a ten-hour trip; he had made the journey to Jordan's several times. But these were not normal circumstances, and neither of them speculated upon what he may encounter. Few words were spoken between them as they worked in tandem to complete preparations for his departure. A small bag was packed, the truck was filled with gas, and a phone call to Jordan's good friend in Joplin gave them a place to stay after they were safely out of Katrina's path.

By 1 P.M. that same day, C.B. was ready to depart. Vanessa knew she could be a great help if she went. An excellent packer and organizer, she was also cool and efficient in a crisis. C.B., however, was more concerned about what unknowns awaited him at Jordan's. He knew his daughter, and he knew something was wrong, so he needed to be unencumbered to handle whatever situation presented itself. There was no doubt in either of their minds that C.B. was driving straight into tumultuous, dangerous circumstances to rescue a strong-willed daughter who may or may not wish to be rescued.

C.B. chose not to confide in Vanessa his greater concern. If they failed to clear the city before Katrina hit, and if they managed to survive the destructive forces of a category five hurricane, what would happen to them when the social structure broke down, and law-and-order became non-existent. With no food, no water, no electricity, and marauding gangs roaming the streets for possibly ten days or longer, myriad unspeakable possibilities loomed. No, he mentally concluded, *I don't want to have to be worrying about Vanessa, too. Jordan will be a handful enough as it is!*

As he began to traverse the southeastern United States, C.B. determined that Interstate 10 would not be a viable route into New Orleans. At Biloxi, Mississippi, he took a small, two-lane road south that brought him to a stop sign where he hesitated a long moment.

Straight ahead rose the angry Gulf of Mexico, surprisingly calm before him in the early evening light, belying the fury of the monster that was gathering her strength from the warm Gulf waters a few miles off shore. To the right was his way into the city, Highway 90. He knew it would be slow-going with street lights, stop signs and traffic, but at least sooner or later, it would get him to his daughter.

On his left was *Beauvoir* in all its grace and beauty. *Ironic,* he contemplated. *Beauvoir means beautiful view. I wonder what it will look like after Katrina has her way."* Originally, a cotton plantation, *Beauvoir* became the residence of Jefferson Davis, the President of the Confederate States. It was here that Davis wrote his memoirs, and this is where he lived out his later years. At some point, a home for Confederate soldiers and their families was built on the grounds. It also became the site for the Tomb of the Unknown Confederate Soldier.

C.B. was thankful that on one of their trips years earlier, he and Vanessa had taken the time to see the original writer's cottage and Davis' writing desk. They would learn that after Katrina, the writer's cottage and about 80% of the other structures on the grounds were turned to sticks of rubble. Miraculously, the mansion itself remained relatively intact. Apparently, the home's six chimneys had held the structure together throughout the raging storm and flood waters.

A honking horn aroused C.B. from his reverie, and he quickly turned the truck toward the center of Katrina's target. Little more than a day later, the very road he now traveled would be pounded to asphalt shards by twelve-to-twenty-foot wind-swept wave surges.

It was after 3:00 AM that fateful Sunday morning by the time C.B. finally reached Jordan's apartment. It was located in Metairie, on the west side of the city near the Causeway and Lake Pontchartrain. Both the apartment and Jordan were a mess. Nothing had been packed, and when the full story came out, C.B. learned that Brent had been incarcerated for possession and drug dealing. This was all alien territory for Jordan.

Her upper-middle-class background had not prepared her for the nightmare into which she had been dragged, and she was overwhelmed. Compound her circumstances with a hurricane barreling down upon her, and, well... it was quickly apparent to her dad that she would never have been able to save herself from the imminent danger looming just outside.

The already-tired father called Vanessa to relay his plans. First, he would sleep, then pick up the U-Haul, pack what they could, and leave by four o'clock that afternoon. What was it that Robert Burns said about the best laid plans of mice and men? *I'll have to reread that poem some day,* he thought later as four o'clock came and went, and they were still packing.

After four hours of sleep, C.B. began his hunt for a trailer. His drive down Airline Highway proved to be a waste of time and gas. When he reached downtown, he found a huge facility with rows and rows of trucks and trailers and carriers of every size and description. Unfortunately, the office was closed and padlocked. He fleetingly considered just taking one, then smiled and decided against being shot for looting a trailer.

He drove further into New Orleans proper where he spotted two parked police cars, so he stopped and asked for suggestions.

"Take the next road to the right and keep on goin'. There's a store down there. It's quite a ways, and I'm not sure it's even open. But if it is, you'll find it across some railroad tracks on your right. It's kinda small, so watch for it."

"Thanks, officer," C.B. responded. "I'm just here to help my daughter get packed and leave."

"Well, buddy, don't wait too long... This thing is big and movin' fast. Good luck!"

With that, C.B. headed down into the lower ninth ward.

He found the U-Haul lot in the 2900 block of St. Claude Avenue, within blocks of both the Mississippi River and the levees that the storm surge, pushed forward by Katrina's force, would top later the next day. When C.B. entered the office, he met Tamika Thompson, a lovely young woman who quickly and efficiently helped him fill out paperwork and rent the trailer he needed.

While they worked together completing the necessary tasks, they talked, and C.B. learned that Tamika's mother was at home caring for her two small children until she arrived. She told him that she didn't know quite what she was going to do or where they would go. C.B.'s obvious look of concern as well as his words of warning must have made an impact on her, because in the middle of their discussion, she suddenly proclaimed that he would be the last customer. Tamika went to the door, locked it, posted the closed sign, then returned to the counter to finish C.B.'s paperwork.

As he left the office, he wished her good luck and offered her the same advice the police officer had offered him, "Get out as quickly as you can, and don't wait too long!"

The parking lot was filled with angry would-be customers who had just been turned away, and, given his audience, C.B., who was normally adept with all things mechanical, fumbled with the hitch and the lights. As soon as he felt reasonably certain the trailer was secure, he hastily left and began his arduous task of working his way back to Jordan. Years later, he would wonder about Ms. Thompson and her little family. He hoped that they had survived Katrina unscathed.

From downtown, C.B. figured Interstate 10 to be the fastest return route, and, for a moment, that seemed to be the case. Then, one by one, the interstate lanes began to come to a halt, and one by one, he began moving across the lanes. By the time he reached the right lane, anxiety had settled in the pit of his stomach, and there was a tingle at the nape of his neck. *What if I become trapped on I-10 like all these other cars? I've gotta get off this freeway at the next exit - no matter what!*

He nudged his pick-up and newly-attached trailer onto the shoulder, squeezed past the cars in the right lane as they rolled to a stop, and drove down an unknown exit into an unknown neighborhood. He wandered, always looking for larger thoroughfares until he found a familiar street and, at last, his way back home.

Katrina's bands of high wind and sheets of rain were already assaulting the city when C.B., Jordan, and Emma, her dog, pulled out of her driveway at 7:15 P.M. that ominous Sunday night. Sadie, her cat, had slipped away during the packing and never returned. Finally, they had to go. They had no choice but to leave Sadie behind.

C.B. wisely decided to head north over the Causeway - a twenty-three-mile expanse over Lake Pontchartrain. As they approached the longest bridge in the world, they could already see the lake water lapping over its banks, and they hesitated. However, police officers were waving them on and shouting, "GO! GO!" So they did.

The wind buffeted the trailer which slowed their speed, but the two or three other cars also making the crossing were not likewise encumbered and were flying past them as fast as the vehicle and its driver could go. They would learn that within hours of their trek across the Causeway, it was damaged, but the shop where C.B. had rented the trailer had been completely swamped by flood waters when the levees were over-crested.

Once they were off the bridge and driving north on Highway 55, they saw cars parked in lines along the shoulder of the road. *They think they're safe.* C.B. thought as he shook his head and drove past. *Better keep on moving. Katrina won't stop at New Orleans!* They rested and had an early breakfast at a diner already packed with others trying to escape nature's fury. When C.B. and Jordan stepped back into the parking lot, he once again felt the "whoosh" of Katrina's winds and rain. At that moment, he realized that even Jackson, Mississippi, was not far enough away and would soon be feeling the great hurricane's wrath, too. Weary though they were, they started the engine and continued their journey. Joplin, to both of them, seemed a very safe place indeed.

§§§

Chapter Four

Logan's Roadhouse, Fairfax, Virginia
Saturday, May 26, 2012, 8:44 PM

Alda was no stranger to the Virginia steakhouse just across the Potomac. When her husband was ashore, the peanut-shelled, classic-rock ambiance was a favorite out-of-the-spotlight venue. The waitress at the receptionist's desk smiled recognition and nodded as Justin pointed toward the hand-waving figure at a back table.

Scott Wasson moved to the chair opposite his and seated Alda, then addressed Justin with a handshake. "Good to see you again so soon, Justin. Very good." He turned to Alda and offered his hand, "I'm pleased to meet you, Mrs. Starkey. And most appreciative that we were able to meet this quickly." Scott signaled to a waiter. "I've ordered several appetizers, and if you like sweet tea, it's good here."

Relaxed confidence for someone so young, thought Alda as she looked across the table, *but hardly the presence of a person considered so preeminent in their field.* As soon as the preliminaries were covered, the Deputy Director came immediately to the point. "Your projections indicate not several decades before irreversible climatic changes begin to occur, but only a few years, not even one decade, before we reach critical points. Please explain why you think your conclusions are more valid than those of the five acknowledged experts in their individual environmental fields, whose findings are the basis of our 2012 Report on the Environment."

The twenty-two-year-old doctoral candidate's reply was just as direct. "G.I., G.O. Garbage in, garbage out. Simply put, my projections are based on a model much more in touch with reality than the one put together for the E.P.A."

Alda leaned back and crossed her arms, but Scott maintained direct eye contact. "Please hear my explanation. Beginning with the basics. Modeling is the development of a mathematical formulation of variables representing a problem, or class of problems, in the real world. Data from the real world is input into the mathematical formula, and results are then processed computationally, which is called numerical analysis. The final product is an approximation which provides a projected solution to the problem.

"As a rule, the simpler the model, the more accurate the results, providing, of course, that all the prerequisite variables are included in the formula. The problems with the E.P.A. model, from my observations, occurred precisely because of the input from the five distinct disciplines."

He rested both elbows on the table, then brought his hands together, sighed and continued his comments.

"Interdisciplinary collaboration has inherent difficulties. I believe your organization uses the term 'intellectual silos' to describe the different terminology and methodologies that vary from one area of scientific work to another.

"You want to know why the E.P.A. projections are incorrect?"

With an edge of exasperation in his voice, Scott moved back and crossed his arms across his chest.

"It's The Trap of the Expert. The specialist in a discipline tends to lose the understanding of the whole in an effort to reveal the parts. It's not a case of connecting the dots, Mrs. Starkey, it's a case of converging the dots. Connecting the dots are two-dimensional, converging the dots are three-dimensional. Variables individually linked may be isolated, linear. Variables overlapping are interacting, reinforcing, non-linear. Simply stated, it's the inclusion of global bifurcation, those small changes in parameters that reach points that cause sudden major shifts in the system."

Justin entered the conversation with a smile. "Global bifurcation. That sounds like something I do after a long night of too much drinking."

Alda nodded demurely while Scott laughed, and continued.

"O.K. Think about water in a tea kettle put on a stove to heat. The water temperature starts at, say, seventy degrees, then slowly increases to eighty, ninety, then higher and higher, until the temperature is over two hundred degrees. The water in the kettle is still water, right? But, a little more heat, just a small amount, to about two hundred twelve degrees, and the kettle begins to simmer. A change is about to happen. Then, with just another small increase in temperature... the water begins to bubble furiously and changes its state, or form, completely to steam. Bifurcation points. Where small changes in parameters, just a few degrees, result in a sudden major shift of a system."

Justin replied with a smile, "Got it. Still fits the situation, though, at the end of a long night."

Wasson joined his smile, barely, for a moment, then continued in a quiet, dejected voice.

"And that's where an outside factor comes into play. The impact of world population growth on the speed of atmospheric imbalance. About seventy-five million people were added to the planet in 2011. To put that in perspective, New York City proper has a population of almost 8.3 million people. So, the world today carries nine more New York Cities of men, women and children than it did one year ago. Or, think of it this way: there are twice as many people on Earth now as when JFK was president. In ten years, we will be adding close to one hundred more New York Cities to stress Earth's natural systems ever so much more.

"So, atmospheric imbalance will not be moving at a speed of simple interest, but, instead, at an acceleration of compound interest. That's why we don't have time measured in tens of years, but quite possibly less than one decade, to avoid an irreversible re-ordering of the earth's environmental characteristics."

Alda's lowered gaze was focused at mid-table as she listened attentively, but Justin caught and followed Scott's glance over the two rows of booths to the bar. The wholesome young woman seated there was slowly shaking her head derisively. Quickly, he turned to Wasson and asked sharply, "Who's that?"

"That," came the answer in a lowered voice, "is Katherine Carson, a Purdue graduate student. Last year, as a reporter for the Purdue Exponent, she interviewed me after I made a presentation at Krannert Auditorium entitled The Sixth Extinction: Toward The Brink?

"We've been here several hours over dinner, discussing whether this meeting was advisable for me to attend. We are in a deeply committed relationship, and she is adamantly against any involvement with you on my part. Unless we are willing to accept a major change in the course of our conversation, I suggest Katy remain where she is for the moment."

Justin looked toward the Deputy Director inquisitively, and Scott Wasson joined the unspoken contact that lasted several seconds.

Finally, Alda replied, "Two questions for you, Mr. Wasson, and then I'd like this young woman to join us. First, if your time frame before critical thresholds were to be accepted as a given, say nine years from now, 2021, please tell me what you see as our options. Be as concise as possible, if you would. "

Scott Wakefield Wasson sat upright. "In terms of human response, one, use technology much more pro-actively to neutralize the effects human activity has already had on the operational balance of this planet, or two, take Stephen Hawking's advice and colonize another planet while we still have the time to do that, or three, be resigned to a very negative effect on many life forms on earth, most notably, mankind. "

"What current efforts do you think should be prioritized and..." Alda's question was abruptly interrupted.

"Pardon me, Mrs. Starkey, but none of them. They simply are not able to reverse what has already taken place, much less be able to stop the ever-increasing impact that humanity will continue to make, even in the short term. Alternative energy is, undoubtedly, the end solution, but what must be implemented immediately is a process that will begin to remove those chemical compounds that we have already caused to become part of our atmosphere."

Wasson pushed forward.

"To obtain that process, we need a political champion, someone to bring our national will and resources to bear, as JFK did for the space program in the 1960s, and we need that person now. We need a Rachel Carson as a catalyst, and a Carl Sagan as a spokesperson, and a NASA-type national technological program imbued with the scientific prowess to achieve what must be done. Our small role, yours and mine, environmental government and non-governmental organizations alike, is basically done. We've revealed the danger, stated the consequences, indicated the solutions. What remains to be accomplished now is beyond us."

"Well then, the last question to you, and please let me know by Monday end, is whether you will join our program, working directly with us in our office? And now, invite Katy to come to our table, if you would."

Katherine Carson left the bar, and moved quickly around the booths and through the tables with the fluidity and grace of a Jaguar closing on prey. Justin began to rise as she swept up to the group with a countenance of disdain and barely subdued malice. "Keep your seat, minion," Katherine commanded, "I don't intend to join your gathering."

She looked down at Alda Martin Starkey.

"I know who you are and what Scott thinks of you, but that makes little difference to me. You're barely a part of the solution, but at least you've had the presence of mind to maintain your integrity, and keep your career in the trenches. I'm sure you could have made the leap into the pit of political appointees, and you deserve credit for not indulging in that degradation of character."

Justin's face whitened and tightened, but Alda raised her left hand in restraint, then turned back to Carson. "Please continue, Katy, if I may call you that."

Katy snapped, "I don't care what you call me. What I care about is Scott Wasson, both personally, and for the singular promise he holds for humankind in the process of self-inflicted destruction. I don't want his head hung low in regret and remorse because he could have been of more value in the struggle to survive. So, you tell me. Is this meeting, the first of this type he's ever agreed to, about to ensnare and muzzle him in some politically-directed, locked-down cubbyhole? I'll tell you what I think. Our government, 'Him the Almighty Power/Hurl'd headlong... /To bottomless perdition, there to dwell.' "

Alda Martin Starkey rose and stood toe-to-toe with Katherine Carson. "Katy," she said gently, "I understand more than you know, and while you have every right to quote John Milton in concern for this person you care for so, listen for a moment to me. First of all, there is only one gifted as he that is seated across from me, in this building, this city, this state, this country, and, quite possibly, in the entire world. I'm very aware of that.

"Before you came over, Scott spoke of the need for a catalyst, and a spokesperson. Perhaps you, or Scott, or a yet unknown will fulfill those roles. As far as expecting political leaders to respond to the greatest challenge humankind has ever encountered, well... let me tell you."

Alda sighed, and started again. "There are few JFKs on the horizon, because it's no longer our elected representatives who determine our national direction. Our politicians are now beholden to a newer, overwhelming power. The top tier of the business world, the corporation, more accurately, the emerged multi-national corporation. A legal, but not human, entity, without moral compass. Behind those brand names we all recognize, there are generally unidentified, faceless individuals who determine humankind's direction, guided only by a profit-and-loss statement. Until we can somehow persuade, or cajole, or downright beg, these powers to modify their business models, perhaps reorganize as 'benefit corporations,' to somehow transfer their assets to sustainable endeavors to reap their profits... oh, please forgive me."

Tears had welled in Katy's eyes, and she spoke up vehemently. "But the natural world is giving such unmistakable signals that are so terrifyingly obvious. Have we become such hot-house flowers that we're isolated from the reality of the world we live in? We leave air-conditioned homes in air-conditioned cars to reach air-conditioned work places, and when the day is done, we reverse the process."

Katy stifled a sob before she continued, "My eighty-one-year-old grandfather in West Plains, a small south-central Missouri town, told me what we now call wildfires, that are almost constant, unending, used to be called forest fires, and were a pretty rare occurrence. And the ongoing droughts, and more floods, and more tornadoes, more violent, Joplin, before that, New Orleans, Katrina that covered almost all of the Gulf at its maximum, the extraordinary melting of glaciers and ice fields, and not just us. Thirty-seven inches of rain in one day in India...

"You know, my grandfather said once that his younger brother was a Marine who was in what was called the 5th M.E.B., Marine Expeditionary Brigade, I think, to the Caribbean. That was fifty years ago, the Cuban Missile Crisis, and it was the only time mankind almost exterminated itself. Until now. What the hell's the matter with our political, our corporate, elite?"

And with that, she walked rapidly away.

Alda addressed Scott as he pushed his chair away from the table, "I'm so sorry. For Katy. For everyone in your generation, and those to come."

"So am I," Wasson reflected, and after he paused, "You can expect me Monday morning," and then went searching for Katherine Carson.

Alda and Justin sat in silence, indeterminately. Finally, he reached for and absently shelled a peanut. "Well, that certainly went... well."

Alda looked up from her reverie. "Actually, I'm pleased. With both of them." A sly smile came over her face as she saw the chagrin in his expression, and with some difficulty, suppressed a deep chuckle.

"Why Justin Lathan Moore, I don't believe I've ever seen you so taken aback. Unprepossessing minion, have the waiter bring us a nightcap with a little more character than sweet tea, and let us take leave of this speakeasy."

§§§

Chapter Five

West Plains, Missouri 2012
"Harvey and Grace"

"**Harvey**," Grace called softly as she gently shook the sickly man sleeping in his worn, faded recliner. "Harvey," She called more loudly this time so that he might hear her over the rhythmic pumping of the oxygen machine positioned next to his chair.

"Eh? What?" he grumbled as he aroused. Between the blaring television and the machines and an occasional grand-baby crying in another room, he could barely hear with what little hearing he had left.

"I have some really special news for you. Are you awake yet? Here, have some tea. I want to tell you something."

Harvey looked at her askance. He did not like being awakened, and it seemed to be happening all the time. The telephone. The doorbell. Babies. What are babies doing in his house? People with awful food. People to take his blood pressure. People to give him a bath. Especially those people! He didn't need a bath. What are those people doing coming into his house hassling him about taking a bath? Why wouldn't people just leave him be.

Grace turned down the TV and sat facing him, tea in hand. "I just received a phone call from Katy, you know, your granddaughter?" She knew that after three heart attacks and several strokes in the last three years, his memory needed a bit of jogging.

"Anyway, she is a research assistant with the E.P.A. this summer. Isn't that great?"

"Huh? She's doin' what? I thought she was in school."

"She is, but not for the summer. She will be in this area, and she wants to come by and see us! What do you think of that?"

"When's she comin'?" he mumbled, and with that, he rolled his head away and went back to sleep.

Grace knew that any further attempts at conversation with Harvey at this point were useless. As she picked up her tea and walked the few steps into the kitchen, she glanced around to assess how much work it would take to spruce up the place. She could see how the little house needed repairs, but there just was no money. Friends and family had helped along the way, but there was always more work to be done. She looked at the kitchen glumly. *Perhaps some fresh paint and new curtains*, she thought. *Can't make it look any worse.*

Gingerly, Grace slid into one of the mismatched chairs and sipped on her tea again. She reminisced about when she and Harvey used to go "boot scootin'" at one of the honky tonks down the road. That was before heart attacks and knee replacements. *Could it really be twenty years ago?* she thought as she shook her head. *Yup,* she smiled, *'fraid so. But, oh, did we have a good time!*

When she met Harvey, he was already sixty-one. She had yet to turn fifty. Somehow, back then, the age difference did not seem to matter at all. They had both had difficult childhoods and kind of rough lives. They didn't hold out much in the way of expectations for themselves or for those around them. When they met at a nightclub, they were both adrift. He thought she was a dish. She thought he was wildly fun.

Not long after, they married and bought a little house in an old neighborhood of houses just like theirs. Bit by bit their aches and pains began to get in the way of their partying until, finally, they could no longer dance, and they stopped going out.

Harvey and Grace did have two rather extravagant indulgences, given their fixed incomes. Harvey loved the old war movies and would sit in front of the TV all day long to watch them, sometimes over and over. To give him more choices, his younger brother C.B. gave Harvey a subscription for Netflix. It was one of the few pleasures he looked forward to.

Grace's passion included her Internet connection with her church friends. They communicated throughout each day discussing their beliefs, particularly as they related to the ever-increasing extreme weather events. Could mankind in his stupidity really be destroying the perfect harmony that God in His Perfect Wisdom created? Furthermore, what could or should they, His children, be doing to make a difference? Above all, she wished to be in His will, for she loved the Lord, and she loved her Internet friends. All their other friends, the party crowd, were either as ill as Grace and Harvey were, or they were already gone.

Grace sighed again, took another sip of tea, and used the table to help herself stand. *It's good to have Katy to look forward to,* she thought as she smiled with delight. *We haven't seen her since...* Grace didn't want to think about the accident that had claimed Katy's parents and Harvey's only child. It was after the funeral that Harvey had his first heart attack, and his health began to fail.

"Poor Katy," she whispered aloud. She sank slowly back into her chair. "Ah, Katy," she said to no one as she turned her tea glass around and around. She struggled to hold back the tears that seemed suddenly to overwhelm her. A lifetime of experience had taught Grace that keeping a tight rein on public displays of emotion was safe, whatever the circumstances, and so she endeavored to do so.

But when Grace began to think about Katy's loss that horrific night four years ago, she began to contemplate her loss, too. A dark, aching loneliness penetrated her soul. For when the teenagers who were racing around the curves and ups and downs of the Hardy Hills plowed into Dennis and Mary's car going eighty miles per hour, Katy's parents were not the only ones to die. Katy's grandfather died that night, too.

The Hardy Hills was a particularly treacherous stretch of road twenty miles south of West Plains and just south of the Missouri-Arkansas state line. Katy's parents had been warned of the steep inclines and sharp curves that made it difficult to drive at night. It was also one of those state roads that had driveways opening directly onto the highway, which meant that vehicles could be pulling out or stopping and turning off the highway at any time. It was not an easy drive under the best of circumstances, and, before they left, the two had promised to return from their Arkansas day trip before dark.

Sometimes the delay of only a few minutes here and there in a life can become catastrophic. So was the case for Dennis and Mary. They left their destination late, knowing that by the time they reached the Hardy Hills, it would be dark. They discussed spending the night in Arkansas, but felt rather foolish and overly cautious for even considering such a thing. And so, despite the concerns and warnings, they began their fateful journey toward home and their precious Katy.

No one else really noticed the change in Harvey. C.B. lived in North Carolina, and by the time he returned for a visit, Harvey had already had the first of his heart attacks. All the changes C.B. saw, he attributed to the attack and Harvey's meds. Grace didn't even try to explain. What good would it do? Garner some sympathy for her? It wouldn't bring back the fun, witty, easy-going man she had loved and enjoyed all these years. She had been looking forward to sharing old age with him.

Now what do I do? she questioned herself wearily. *I'm not even sixty-five.*

Well, for one thing, she chastised herself, *you stop feeling sorry for yourself and take the best care you can of him. Grouchy or not, he's still a good man! He never promised that life with him would be easy.*

All things considered, she finally concluded, *we've had a pretty good life together – probably better than either one of us expected for ourselves. I know he didn't want this for me, any more than I wanted all that he has gone through for him. Life just has a way of not always going in the direction we want or expect it to. What do we do? Just sit down and quit? Well, maybe some folks can, but that sure isn't me.*

Katy went back to college almost immediately after the funeral and had not been back to Missouri, even for a short visit. Grace understood how difficult it would be for her to return. Katy was the only child of an only child, and they were a tight-knit little family. Grace had been around since Katy was born, but there was such an extraordinary bond among the three of them that even Grace felt left out.

Harvey had utterly doted on Katy, and she adored him. She would sit at his knee and listen endlessly as he told of his many adventures, some real, some exaggerated, some pure figments of his imagination. It made no difference to Katy. He would take her to the Saturday cartoon matinée at the theater or to the park to feed the ducks and play on the swings. Harvey taught her to read and to ride a horse, and as she grew, so did their conversations. He insisted that she be an educated woman and an independent thinker. He believed in her, and she thrived in his sunshine.

I have tried so many times to prepare her for the changes here, but I don't think she hears me. As far as that goes, there's just nothing more I can do.

"It will be good to see her again." Grace sighed absently as she glanced at the clock above the sink. Suddenly, she was jolted back to reality. Her life revolved around Harvey's daily routines with little time left over for her wants or needs. If she expected to return in time for his next round, she would have to go quickly. She once again helped herself stand.

Oh, well, she thought as she surveyed the kitchen, her chipper mood returning. *I need a project about now, and this one I am going to enjoy!* With that, she grabbed her keys and went out in search of just the right color of paint and just the right fabric for the curtains.

§§§

Chapter Six

Environmental Protection Agency, Washington, D. C.
Friday, June 15, 2012, 10:03 AM

The Deputy Director of the Air, Climate, and Energy Research Program rose and addressed those at the conference table in her office. It was a small group, and the table had been set for six, with Alda at the head.

"Now that we have refreshed ourselves, and our final points have been covered, I have several comments for all seated here. Over the last several weeks, we have had many discussions, our decisions have been made, and the final arrangements are complete.

"We are sending forth from this office a person highly esteemed in the scientific community. The assignment he and his volunteer assistant have accepted are fraught with unknowns and unpredictable implications, and I know each of us holds fervent hope their endeavors will be fruitful. We heartily wish them Godspeed."

After an abbreviated, subdued applause, she continued.

"Fill your glasses and get comfortable, because what I have to say will require some time and your full attention. We must be aware of, and prepared for, the potential ramifications of the actions we are setting in motion today. Each of us may well be impacted on a personal basis in unforeseen ways by that which transpires over the next several months. I do not wish to be an alarmist, but forewarned, as is said, is forearmed."

Alda pointed toward her Executive Secretary at the opposite end of the table, and the four others looked in her direction.

"Patty McCarty, you and I go back some time together, and it is you I am most concerned about. You and John still have children at home.

"Should it become necessary, you will have an on-par position available with one who shall remain unnamed at this time, and I feel certain your competence will be valued as much there as it has been here."

She nodded to the left toward her Administrative Assistant.

"Carter, at twenty-five, your future is bright. The three years you've been with the Agency have been illustrious, and I'm very pleased that your time has been spent in my office. As events unfold this summer, if you think it prudent, do not hesitate to distance yourself from me. You've done well for a beginning in a small South Carolina town, and, in large part, you can thank your parents in Bishopville for where you stand today. If I were here long enough, which I don't wish to be, I would probably be reporting to you. Take care, Carter, that you rise morally intact."

The Deputy Director turned her attention across the table to the right and looked intently at Justin Moore. As she addressed her Research Resource, Katherine Carson, seated between the two, noticed the tenderness in her eyes, and more, a softness, almost a caress.

"Justin, the first time I saw you, you were less than a week old. Your father was then a young, recently-arrived environmental scientist in the Agency, and he and your mother were instrumental in my recovery from a difficult time in my life. It was your mother who introduced me to Bill Starkey, a naval officer fresh from Annapolis, who indeed proved to be fresh. You have been raised in the Agency, and I believe those experiences imbued you with out-of-the-box thinking and a deep spirit of adventure.

Richard Moore brought you with him at every permissible opportunity, and on some occasions that weren't, but conditions were different in the 1980s, more tolerant, less formal, less security conscious. You are very well known in this organization, and would be highly sought after if you were available, should that occasion arise."

She noticed the almost palpable concern expressed on the faces around her, and softened her voice.

"Scott and Katy's reports must have the credibility of the E.P.A. seal to achieve optimum effect. Therefore, this office will function as the conduit to the media of Scott Wakefield Wasson's dispatches on the State of the American Heartland. Not if, but when, the fossil fuel industry and their governmental proponents learn of the source of these releases, overpowering forces will be brought to bear on the Agency, and that weight will inevitably come to rest upon the responsible entity.

"Let me assure each of you, I am prepared for that eventuality. I neither expect nor desire to attain a higher level than this position I now hold, and, truth be told, twenty-six years is long enough to be in this organization.

"Earlier this year, my fresh naval officer reached one star rank, rear-admiral, lower half, comparable to a brigadier general in the army or air force. I foresee no great hardship should the aforesaid scenario come to be reality. There also exists the possibility that an environmental, non-governmental organization might find some value left in a discarded relic such as myself."

Alda went silent for a long moment, then moved from the head of the table to her right, and stood directly behind Katherine with her hands on the back of Carson's chair.

"Rachel Carson has been referred to as the mother of the modern environmental movement, and, though she died in 1964, the creation of the Environmental Protection Agency was undoubtedly due in great part to her environmental activism. In 1986, when I was about your age, Katy, I came to the Agency in Reagan's second term.

"At that time, the Agency was still rebuilding after the devastation wreaked by Reagan's first E.P.A. Administrator, Anne Gorsuch. She reflected Ronald Reagan's attitude toward the environment as typified by his observation that 'approximately 80% of our air pollution stems from hydrocarbons released by vegetation, so let's not go overboard in setting and enforcing tough emission standards from man-made sources.' Her most lasting legacy was 'making anti-environmentalism one of the ten commandments of being a Republican.'

"In 1989, Bill Reilly, a true environmentalist, became the E.P.A. Administrator, and my enthusiasm for, and decision to make a career in the Agency developed. Then came Carol Browner, the longest-serving administrator of the E.P.A. She developed a cleanup program for the *Brownfields*, the abandoned, contaminated properties that blighted the American landscape, and she earned my full admiration for that achievement alone. When the Republicans gained control of Congress in 1994 with the highly touted Contract with America, Browner successfully fought their attempts to weaken the Clean Water Act. She advocated inclusion in the Kyoto Accords that Clinton signed representing the United States in 1998. That was a heady moment, but before it was ratified, George W. Bush rejected it, and to this day, the United States of America is not part of the treaty."

A sense of sadness and disappointment settled on the Deputy Director.

"Shortly before the turn of the Millennium, my mostly modest achievements resulted in an appointment to the position I now hold, but my joy was short-lived. Dark times quickly ensued for the Agency.

"Intense political pressure pushed in on the Environmental Protection Agency from the highest levels of the new Administration. During the following eight years, we had three Administrators, Christine Todd Whitman,

Republican governor from New Jersey, Utah's Republican governor Michael O. Leavitt, and an Agency veteran of twenty-four years, Stephen L. Johnson.

"At the beginning of her tenure, Whitman commissioned a report on the current status of the environment, based on the latest scientific findings. The White House removed evidence from the report that documented the rise of global temperatures and the contributing factors related to human activity, which not the least, of course, was the fossil fuel-generated atmospheric pollution. After almost two-and-a-half years, she finally resigned. That was at the end of June, 2003. Mr. Leavitt, as did our President at the time, felt pollution control would better be dealt with by the individual states rather than the federally-mandated Environmental Protection Agency, and furthermore, suggested voluntary environmental protections would be preferable to mandatory compliances. It was a relief when one year and one month later, he was appointed to head the Department of Health and Human Services."

Alda Martin Starkey's countenance turned stony.

"Then came Stephen L. Johnson, a bureaucrat with twenty-four years in the Agency. Johnson began well, but within a year, the Administration had found a way to bend him to their political will, and to the agenda of those interests they represented. Conditions within the Agency worsened continually, until, at the end of that Administration, we were operationally bouncing along on rock bottom.

"Our plummet to the depths was noticed at many levels, and in the entire country as well. On December 11, 2006, the Union of Concerned Scientists issued a statement signed by 10,600 leading scientists including Nobel Laureates, that called for the restoration of scientific integrity to federal policy-making.

"The Philadelphia Inquirer published an investigative report in a four-part series entitled "Smoke and Mirrors," and in your briefing folders, there is a reprint of the first of the articles. An Internet search will, no doubt, provide you with the full content of the reports."

An Eroding Mission at EPA

The Bush administration has weakened the agency charged with safeguarding health and the environment.

December 07, 2008 | By John Shiffman and John Sullivan, Inquirer Staff Writers

WASHINGTON - On Dec. 5, 2007, EPA Administrator Stephen L. Johnson prepared to send the White House an extraordinary document. It declared that climate change imperiled the public welfare - a decision that would trigger the nation's first mandatory global-warming regulations.

Johnson, a career scientist, knew that his draft would meet with resistance from anti-regulatory ideologues at the White House, but he believed the science was solid.

According to confidential records reviewed by The Inquirer, Johnson cited strong evidence: rises in sea level, extreme hot and cold days, eco-system changes, melting glaciers, and more. Minor doubts about long-term effects, he wrote, were not enough to alter his conclusion. Two sentences in Johnson's draft stood out. In sum: The U. S. emits more greenhouse gases from cars than most countries do from all pollution sources. This fact is so compelling that it alone supports the Administrator's finding.

At 2:10 P.M., Associate Deputy Administrator Jason Burnett e-mailed the climate-change draft to the White House as an attachment.

What happened next became Johnson's defining moment and cemented President Bush's environmental legacy, serving as the low-water mark of a tumultuous era that has left the E.P.A. badly wounded, largely demoralized and, in many ways, emasculated.

White House aides - who had long resisted mandatory regulations as a way to address climate change - knew the gist of what Johnson's finding would be, Burnett said. They also knew that once they opened the attachment, it would become a public record, making it too controversial to rescind. So they didn't open it.

They called Johnson and asked him to take it back.

The law clearly stated that the final decision was the E.P.A. Administrator's, not Bush's. Johnson initially resisted - something Burnett admired - but ultimately did as he was told.

Outraged, Burnett resigned.

In July, Johnson issued a new, censored version, a pale imitation of the original climate-change document. The old muscular language - including key sentences about U.S. car emissions and the irrelevance of any lingering doubt - was gone. Most of all, the new document no longer declared global warming a danger to public welfare. The move effectively postponed any strong action on climate change well into the next administration.

§

Alda Martin Starkey returned to the head of the table, changed tone, and delivered the charge.

"Scott and Katherine, you two are the tip of our arrow. You will be in bulls-eye flight this very afternoon, aimed to diminish extensively any lingering uncertainly left in the American public, as they continue to experience increasingly unusual weather changes in their everyday lives. That lingering uncertainly is promoted relentlessly by the fossil fuel industries, and advanced by their proponents in the Congress of the United States of America, even in the presence of overwhelming, rationally undeniable, evidence to the contrary.

"Like the tobacco industry before them, the fossil fuel industry will use every tactic to foster confusion and keep doubt in the minds of our citizens. Now, however, the advent of extreme weather conditions has resulted in a growing public opinion that something, indeed, is amiss."

Alda made full eye contact with each, one by one, around the conference table.

"Our people, particularly in the American heartland, have already begun to feel the effects of increasingly unusual weather. When Katrina gained category five force from the warmed ocean waters, the swirl covered virtually the entire Gulf region from Florida to Texas, and that was seven years ago. If you don't remember, go back and find the archived images. The unprecedented Joplin tornado, on the ground for over six miles with a swath at least three thousand feet wide and winds in excess of 200 miles per hour, was an example of the increased frequency of extra-violent wind storms.

"The same for summers and winters. Now it's not unusual for Texans to endure summer month-long, triple-digit temperatures. On the other hand, winters as far north as Illinois have almost no snowfall, and I assure you, it wasn't that way fifty years ago. Last year, the Mississippi flooded. This year, it dropped twelve feet in places, and river traffic has gone aground. Nebraska down through Kansas and Oklahoma and Texas is a thousand-mile-plain of severe drought."

Carter spoke up unexpectedly, low-voiced, " 'I will show you fear in a handful of dust.' "

Katy responded, "From <u>The Waste Land</u>."

Alda nodded, "T. S. Eliot, yes. All this and more is in our faces now, before our very eyes. We, the people, in the end, will determine our destiny in this situation. The American public cannot depend on

Congress for leadership, not because the structure of our representative form of government is broken, but because it is corrupted. Too many politicians are far too responsive to the monstrous-moneyed fossil fuel industry that willingly provides the campaign funds considered so necessary for political survival.

"As a direct result, our federal government provides four billion dollars in subsidies to this industry annually, while, outrageously, the merged-to-a-few giant oil companies record bloated profits on a continual basis. The hose at the gas station has drained the wealth of the middle class like the proboscis of a West Nile virus mosquito.

"There is a similarity between this situation and the power that the tobacco industry once exerted over America. Twenty or thirty or forty years ago, most people smoked tobacco, evidenced by the cigarette butts that littered the ground everywhere. Even when the Surgeon General of the United States made it abundantly clear, in no uncertain terms, that smoking caused cancer, the tobacco industry continued unabated, and actually began including highly addictive additives to cigarettes. Now, those once ubiquitous cigarette butts are rarely to be seen. It wasn't the politicians who lead the way as the influence of those financially-motivated parasites on the health of our people was minimized, and it won't be the politicians who lead us now, either."

Alda paused, sipped water, then looked to her immediate left and right.

"Scott, I will inform Region Seven headquarters in Kansas City and Region Six headquarters in Dallas late this Friday afternoon that you and a volunteer assistant will be in their areas for several weeks conducting independent 'eyes-on' assessments for the Air, Climate and Energy Research Program. Your credentials will identify you as an E.P.A. Independent Field Assessment Officer, reporting directly, and accountable only, to the A.C.E. Deputy Director. As an I.F.A.O., you will be issued an unlimited amount debit card for all expenses you deem necessary. This card will require a double P.I.N. entry, and you must personally confirm your possession of the card every twenty-four hours.

"You will not be required to check in with any E.P.A. facility, and I would prefer that you don't. Use William Least Heat-Moon's approach, and <u>Blue Highway</u> your travels from one point to the next. As we have decided, Kansas City International will be your first flight destination, and the status of

Joplin, Missouri, one year after their EF-5 tornado is your first assignment.

"We know New Orleans is nowhere near pre-Katrina conditions seven years after the fact. It is vitally important that we become knowledgeable regarding the process and time required for American cities to recover from weather-related catastrophes, simply because there is more of the same coming to the American people, especially in the Heartland.

"Katherine, you have been accepted as an E.P.A. Volunteer. Your responsibility is to assist Mr. Wasson as needed in the performance of his assigned duties. Your position includes a mandatory requirement that absolutely no Agency funds be expended in any fashion for any type of expense incurred by you. However, we do have sponsorships available from parties unaffiliated with the E.P.A. that will cover your expenses in full. Carter has completed packages for both of you, with detailed explanations.

"By the way, Justin has ascertained that you do have a relationship to the legendary Rachel Carson. The father of your grandfather in West Plains, Missouri, was a first cousin who spent time during childhood summer vacations with Rachel Carson.

Technically, that would make you a first cousin, thrice removed, I believe.

"Katy, you would do well to keep in mind that earnest, high-profiled activism for valid causes is admirable at any age, but much easier to engage in when you are young and unencumbered. If you have those esteemed inclinations, which apparently you do, enjoy the freedom inherent in this period of your life. Marriage, children, mortgages and other adult life responsibilities tend strongly to constrain 'damn the torpedoes, full steam ahead' behavior."

Alda Martin Starkey lifted her chin, and spoke with resolution.

"Now, however, I have come full circle. Those concerns are no longer relevant in my life. Your example of demonstrated conviction and commitment has rejuvenated my dedication and desire to step firmly to the right side of history."

The Deputy Director squared her shoulders, sighed deeply, and leaned to the right to peer directly at Katherine.

"There is an issue with you that must be resolved, and now is the time to do so. This is a town rife with temptations of many sorts, young lady, and I have been informed that you have succumbed to one. One that may appear somewhat minor at this moment, but..."

Katy's eyes widened, and she inhaled deeply through parted lips.

Projecting stern severity, Alda continued. "Several weeks ago, you offended the young man to your right, and he has kept an eye on you since then."

Justin's smile seemed mocking as Katy glanced toward him.

"Katherine, the White House is on the far side of this block to the west, and the Congress of the United States is twelve blocks to the east. Between here and there, it appears you have fallen, repeatedly. In fact, just east across 12th Street. There, in but a few paces, you have evidently indulged yourself decadently."

Everyone at the conference table nodded knowingly, then smiled, and finally, laughter reigned.

"Yes, Katherine, the path you have worn to the Ben and Jerry's located there has indeed been noticed. And Katy, there is a lesson here for you."

Alda Martin Starkey, with amusement dancing in her eyes, commented sassily as she walked away from the table. "Revenge is best by a dish served cold. Frozen, actually, in this case."

§§§

Chapter Seven

Arlington, Texas 2012
"Nick's Dilemma"

Nick smiled as he hung up the phone. *At least now I can talk to her without getting angry every time*, he mused as he slogged his way home through Dallas traffic and another day of relentless triple-digit heat. Nick and his mom had finally agreed to work at rebuilding their relationship after a tumultuous second marriage had all but torn both of them apart. In the end, Edward died of malignant melanoma, and Vanessa married C.B., an old college sweetheart. When Vanessa and C.B. moved to North Carolina, only the painful memories remained, which neither mother nor son wished to remember. Nick had been skeptical about the new arrangement he had made with his mom, but liked how things were going so far. *At least for once*, he thought, *she agrees with me.*

Nick and Kara had been talking about the imbalanced atmosphere and its impact on their lives for so long that detailed discussions and explanations were seldom necessary. Parts per million. Tipping point. Pollutants. Ozone. All were painfully real to them as they watched their precious Ashley struggle daily for each breath. Dust storms were increasingly blowing what was once fertile west Texas top soil through the metroplex. Brush fires were everywhere, continually depositing soot and ash on the hoods of cars and in Ashley's sandbox. Even the night gave minimal relief from the oppressive heat. Without a word, both knew that their time in the metroplex was running out.

It was difficult to think about leaving. Both had grown up in the area, Kara attending choice Arlington schools, and Nick spending all of his growing-up years studying at a prestigious private school in Dallas. After graduating from the University of Texas, Nick returned to the metroplex where he was introduced to a much-younger Kara who, with her quick wit and sassy charm, absolutely mesmerized him. He had met his match in every way. Intelligent, perceptive, sensitive, attuned to the world stage. There was little that he needed to explain to her. She was also independent, strong-willed, and just as bull-headed as he. Yes, he had met his match.

Nick knew that he could never convince her to leave her parents behind when the time came that survival dictated moving. He also suspected that her parents would never leave, regardless of how convincing he might be of the peril for them if they were to remain. *Please, Lord,* he would pray silently each day. *Give me the strength and courage to save Ashley.* He never verbalized the rest of the thought. He refused to consider what that might entail.

Nick liked his professional environmentalist position with the state of Texas. His area of responsibility encompassed the entire Fort Worth/Dallas region and included working on political as well as environmental issues. Not only was he surprisingly adept in the political arena, but also, he was introduced to myriad other professionals in his field and related fields at seminars and conferences throughout the region.

Various geographers, cartographers, statisticians, and scientists would arrive ostensibly to discuss conservation or impact studies, but somehow, the conversations always reverted to the primary topic on everyone's mind – climate change and its impact on their communities. What is the impact of the imbalanced atmosphere going to be on the maps that currently exist? How dramatically and how rapidly are changes beginning to take place?

The ArcGIS technology that is in use in most government offices throughout the country had already been used to show some predicted potential shore line outcomes given current global warming data. Is it capable of handling the magnitude of change that is about to inundate the system?

At a conference in July, Nick and his close circle of colleagues met for their private updates. The task of each member of the personally-created group was to gather whatever pertinent information he or she could glean from his or her respective discipline about the latest on atmospheric imbalance as it pertained to that discipline, then report back to the group at their next meeting. They had been collecting data for years, but the growing sense of urgency and anxiety was new, and their meetings were becoming more frequent.

"Are you considering a move out of the metroplex, Nick?" Randy, the cartographer, asked.

"Not yet," Nick responded slowly and shook his head. "Kara still has two years to go on her accounting degree and CPA certification. We both want her to finish, but I don't know if we have that much time. All the data just keeps looking worse every time we meet." He shrugged his shoulders and sighed.

"Yeh, I know what ya mean," Randy nodded. "We have relatives up in New Hampshire, and we're already talkin' to them about movin' up there. It's tough to know when to leave, but... Nick, don't wait too long!"

Nick left their meeting deep in thought. All the input, the data, the polar ice sheets, Randy's warning, tumbled over and over in his mind.

"Randy is right," he said aloud to himself as he wended his way homeward. "This is the time for us to make our preparations to move, while the possibility of jobs in the mountains still exists, and before the rest of the millions who will be affected decide to make their mass exodus from America's heartland."

He knew the coming conversation with Kara would be difficult, but they still had some time. Besides, the West Nile virus crisis this summer was all around them. Some of the deaths were surprisingly close by, and when the papers revealed that more than one in three of the mosquitoes caught in the CDC traps was infected with the virus, Kara all but refused to let Ashley out of the house for weeks. *Perhaps now,* he hoped, *she will be more amenable to my plans.*

Kara intensely dreaded their discussions after his meetings. It was always *MOVE, MOVE, MOVE.* She had lived in the metroplex all of her life. This was her home. She was, after all, a real born-and-bred Texan, and whether anyone else knew or understood the concept, there truly was something special about Texas. Even if the discussion had not included leaving without her parents, which it inevitably did, the thought of being somewhere other than Texas was an anathema to her.

Her parents were another major issue. How could she leave them behind when she knew what lay ahead for the metroplex. Her fifty-six-year-old father had been unemployed for several years before finally finding a decent position with an engineering firm. He felt certain that if he left that position, he probably would never work again. Silently, she and Nick had both agreed with him. Her mother, on the other hand, took medical dictation in a doctor's office, so her prospects for employment elsewhere might be better.

When Nick talked to them about the imminent danger the impact of the imbalanced atmosphere posed, Kara's parents were dubious. In that respect, they were not alone. So many confusing and misleading statements had been published over the years that they had achieved their intended goal: To muddle and befuddle the issue in the minds of citizens to the point that people would not complain, and, consequently, nothing would be done to change the habits of those major polluting industries.

Her parents could not imagine that all of the dramatic events which Nick described so vividly would be occurring within the next ten years. Yes, they had seen for themselves the drought, the brush fires, the dust storms, and the high temperatures, but they had seen them all before. This was, after all, Texas. They did, however, have to admit that the West Nile virus and all of the people dying from mosquito bites was new, and more than a little frightening.

They also reminded him of how worried everyone was, Nick's mother included, about the Avian flu and a world pandemic. "Stock up for three months," everyone said. "Be prepared to defend your home," everyone said, and nothing ever came of it. They did not want to sell their home and leave everything behind, only to find out that *If* anything is ever going to happen, it might be hundreds of years from now.

Kara knew all of their arguments *and* all of his. How was she ever going to reconcile the two?

§§§

Chapter Eight

Video Conference, Joplin, MO - Washington, D. C.
Monday, June 18, 2012, 9:02 AM

The Deputy Director of the Air, Climate, and Energy Research Program was flanked at her desk by Carter and Justin. Scott Wasson had connected for their first video conference since his and Katherine Carson's departure the previous Friday afternoon.

"Good morning, Mrs. Starkey, and Carter and Justin, too, I see."

After a chorus of greetings, Alda began, "I trust your flight went well. Any problems to date?"

"None so far. We rented a forty-foot motor home, wound our way through Kansas City into Missouri, and drove down Highway 71 to Joplin. Easy drive, actually, straight south just over a hundred miles, running parallel to the Kansas-Missouri line. The motor home is newer, nice. We drove around Joplin Friday afternoon and evening, just for a first look. Saturday and yesterday were spent getting a basic feel for the town. Fair sized, about fifty thousand people. At night we've parked at the new Walmart Supercenter, along with several other R.V.s of various types and sizes.

After a pause, Alda asked, "The motor home. Nice, you say?"

"Yes, quite."

"Good. Full kitchen?"

"Yes, and dining area."

"Good. Adequate bath?"

"Yes, shower and jetted tub."

"Good. And sleeping berths fore and aft?"

"Uhhh... yes."

"Good, very good."

After an awkward moment, as Carter rubbed the tip of his nose with an open hand to hide his smile, Justin outright snorted and injected, "Well then, give Katy our greetings when she's up and about. Any other questions regarding the motor home, Deputy Director?"

"No, no, not really," Alda replied rather hastily.

Then, more formally, "but I would like your first impressions of Joplin."

"What you would expect to find in the heartland, I think. Industrious people, recovering from an extraordinary devastation. The debris in the swath of destruction has been hauled away, a tremendous undertaking in itself. It's clean, but... barren. The businesses have rebuilt, to a great extent. The residential areas are a different story. Surreal, literally."

The Deputy Director rubbed her forehead with her fingertips, "And this, a year and a month later. Is this what can be expected, realistically, as our people endure increased extreme weather events?

"We need a comprehensive report, Scott, every aspect. Stay two weeks in Joplin, gather your information, have Katy generate a fully-detailed dispatch, countersign your concurrence, and overnight it to us no later than Friday afternoon, June 29th. The following Monday, July 2nd, at this time in the morning, I will review your report with you, and discuss your findings as necessary."

Alda Martin Starkey swiveled away from her desk, rose, and walked to the windows to gaze once again toward the Hill. *I seem to be doing this more and more often*, ran through her mind, and it was the same thought shared in the glance between Carter and Justin. She returned to her desk, and addressed Scott, "Thank you for the indulgence of a quiet moment."

Before Alda could continue, Scott turned his iPhone just enough to include Katherine's image. "Katy was up early to make a Walmart run before the crowds began in earnest, and she's back. Would you like a word with her?"

"Of course. Good morning, Katy."

"Good morning, Mrs. Starkey," then nodded to either side of Alda, "Carter, Minion."

The ghost of a smile flitted across the Deputy Director's face as she asked, "So, a stocking of staples for the homestead-on-wheels?"

Katy laughed, "Yes, and a confirmation of American cultural continuity, as well. Scott, you remember that Walmart across Northwestern Avenue from the Purdue campus, and the irritating maneuvers required when shoppers left their carts askew in the middle of the aisle to wander for who knows what? Most, I must admit, were left by those of my gender. Truth be told, they don't miss a beat in Joplin, Missouri, either."

Alda's smile widened, "Tell me, have you found motor home travel acceptable so far?"

Katy was enthusiastic.

"For a vehicle forty feet long, the ride is stable, solid, and it corners without sway and roll. We like it. The kitchen is well-laid out, everything is handy. The bath is fine, good shower head, and a relaxing jetted tub. And the whole back end is all bedroom, king-sized, really comfortable. I feel just great this morning!"

With quick waves, Carter and Justin left Alda Martin Starkey's side simultaneously, and headed for the door.

"You can hardly imagine... how pleased... I am to hear that, Katy." Then a little too quickly, "Well, we must go. Travel with care, and be... safe. Talk with you later."

Alda leaned forward, broke the connection, and watched with narrowed eyes as Carter and Justin wrapped their arms around each other's shoulders and staggered, chortling, out of her office.

§§§

Chapter Nine

Joplin 2011
"Jordan's Near Miss"

All Jordan could see when she looked around was utter devastation. The overwhelming sense of grief and loss and pain that she shared with this old but caring town brought her to her knees, and she wept. The town and its people had been a light in the darkness to aid her in finding her own path and a soothing balm to help her heal after the nightmare of New Orleans, and to see such death and destruction descend upon it and them wounded her very soul.

Jordan's move to Joplin after Katrina gave her the opportunity to pursue once again those goals she had so long ago abandoned, including going back to school. She was good at bookkeeping, which helped her to secure a position in the procurement office of Missouri Southern State University. Being on campus daily also encouraged her class attendance and increased study time. Jordan was recognized as a diligent worker and conscientious student. Thus, when she walked into her office that Monday morning, her boss had no qualms about responding positively when she suggested that they all be allowed to leave so they could help the town in some way.

Mr. Williams called the staff together to tell them of his decision to dismiss them all. He concluded by saying, "Now go. Help where you can. Our friends and neighbors need you more than this office does. But please, be careful. There is danger everywhere. I have just heard a warning about St. John's Hospital. Apparently, it suffered a direct hit. It's awful!"

He paused a moment, lowered his head, shook it slowly side to side, and said a silent prayer. Then he continued, "The fire department and haz-mat people are concerned about possible explosions emanating from the hospital wreckage and are trying to keep a wide perimeter around it.

"If you can change into heavy boots or shoes, thick jeans and leather gloves, do so. Also, there are some pharmacies who were not hit that are opening their doors and giving tetanus shots to anyone who comes by. My understanding is that they will be offering them in the field by tomorrow. But, if you're going into the war zone, get one before you go. You will need it. You can't walk five feet without being poked or scratched or stabbed by something.

"You will still get paid for this week, so don't worry about that. It's the least the University can do for this community. God bless you for helping our ravaged little city. Now, go, and be careful!"

The morning after the tornado had swept through continued to be a threatening, dangerous weather time. The storms had not completely cleared out of the area, yet people were still trapped in the rubble of flattened homes and buildings, and those who were not directly impacted by the initial catastrophe were selflessly exposed to the danger in the midst of the chaos as they tried to help their neighbors. One rescue worker, a police officer, was struck by lightning during that first day and died as a result.

Ironically, Jordan's home was a few blocks away from the point where the vortexes joined and the tornado began its deadly six-mile swath of the city, yet she never saw it. She knew there was supposed to be a severe storm that night, but such a forecast had not frightened her. After all, she had been in severe storms before. She was not even surprised when the electricity was out all night.

There were things, however, that she did find odd. She heard the ferocious wind and expected to see the trees blowing in all directions, but they weren't. When she looked out, she was surprised to see that they were all leaning in the *same* direction.

Curious, she thought, *I've never seen that before.* She heard things banging on the walls but saw nothing flying around.

All night long, she kept hearing sirens and helicopters and receiving text messages in those moments when her phone did work saying, "Are you OK?" and "Did you survive?" *This is crazy! What is going on?* She wondered. Of course, without electricity, Jordan had no television or radio and had no idea what she would stumble upon when she arrived on the Missouri Southern campus that horrific Monday morning. It seems that the whole nation knew about the tornado before she did, and it missed her by only six blocks.

§§§

Chapter Ten

Video Conference, Joplin, MO - Washington, D. C.
Monday, July 2, 2012, 8:57 AM
"State of the Heartland: Joplin, Missouri"

Alda Martin Starkey was in a subdued mood as Scott Wasson's image focused and cleared. His report lay spread across the top center of her desk, and several pages had been pulled lower at various points, each with red-circled passages scattered from top to bottom. She thumbed her chin as her gaze swept slowly back and forth over the documents, then, despondently, leaned back and spoke to Scott.

"Sunday afternoon generally affords an hour or two of discretionary time when I'm without spouse on shore, and, occasionally, I'll come here to digest significant information in quiet solitude. Yesterday afternoon and well into the evening came and went at this desk without my notice. All of this is very distressing. The death and destruction, human agony and anguish, the pace of recovery, and your conclusions."

The Deputy Director closed her eyes, pinched, then massaged the bridge of her nose, and came upright.

"An excellent presentation, Scott, and well-crafted. My compliments to you and to Katy. Given the quality of this report, so well detailed and documented, there is no question but that you should generate a report concerning the Mississippi River Basin from Cairo to New Orleans. As planned, pick up Highway 60 south of Joplin in Neosho, which will take you west to east across southern Missouri. You might logically consider an overnight stop at mid-point, perhaps West Plains. The following day will put you just west of the Missouri-Kentucky border, the area of the major flooding in April, 2011. Please let me know when you get to that point.

"So," Alda continued, "this is the first of the 'eyes-on' reports on the heartland. That wide strip from the open Canadian north to the Gulf of Mexico on the south, bordered by two extended mountain ranges on the east and the west, has always been subjected to major storm activity. Kansas, Oklahoma and Missouri have been called 'tornado alley' for good reason.

"But, tornadoes of this magnitude..." she was shaking her head, "extremely exceptional. Everyone with a mind to do so can easily research the incidence of increasingly severe, record-setting weather events in the Midwest and Great Plains states over the last fifteen years."

She paused, then pointed to her desk.

"So simply and concisely stated, your description of weather factors, their interplays, the impact on fifty thousand men, women, and children in one short half-hour, a small city's recovery requiring month upon month of effort and expenditures, now over a full year and yet open-ended, and the inescapable conclusions."

She leaned forward and retrieved a page at the left of the spread.

"Here, at the beginning, cold air fronting down from the north across Canada, seasonally typical. Dry winds from the west, from the lee side of the Rockies, over Kansas and Oklahoma, but much drier than usual, more moisture pulled out of the air by the land in drought beneath. Warm air, exceptionally moisture-laden, bulging up from the Gulf, warmer than normal from the higher-than-average sea water temperatures. In an area of northeastern Oklahoma and southeastern Kansas, they all coalesce, mix, swirl, and angrily present Joplin, Missouri, with the extraordinary, super-celled, multi-vortex result. Straightforward, ergo pepto bismal, process.

"Six miles, 10,500 yards, on the ground. So that's 105 football fields laid end-to-end. Three-quarters of a mile wide, 1,320 yards, thirteen football fields wide. Total encompassed area of destruction, four-and-a-half square miles, almost 8,000 square yards, approaching 25,000 square feet, all subjected to wind forces of two hundred and more miles per hour."

Alda stood, and reached for another selected page.

"Time frame, estimated, 5:34 PM to 6:12 PM, approximately thirty minutes. Over 150 dead. Five men, women or children killed for every minute in the EF-3 to EF-5 impacted area. One death every twelve seconds. Joplin's newspaper, the Globe, reported that half died in their residences, a third died in non-residential areas, and the remainder died in vehicles or outdoors. Almost 1,000 injured, thirty casualties for every minute in its on-the-ground path. One injury, some horrific, every two seconds.

"You also included this excerpt from The Kansas City Star, 'Joplin still bears scars of deadly tornado,' by Eric Adler (Posted on Sat, Dec. 17, 2011 11:54 PM), [read more here: http://www. kansascity. Com/2011/12/17/ 3 326113/joplin-still-bears-scars-of-deadly. Html #storylink=cpy]

' "What that night brought is hard to express in words," said Allen Overturf, 39, a nurse who worked the night of the tornado as director of critical care services for Freeman Health System, which treated the mass of

Joplin's wounded. "We saw hundreds and hundreds and hundreds of patients with massive trauma - exposed brains, exposed organs, very broken bones, unlike anything I ever could have perceived." [1]

The Deputy Director's voice quavered, and she stopped. After a long moment, she held another of the pages between her face and Scott's image, and continued.

"Within the area impacted by the tornado, almost 7,000 homes were totally destroyed, left as piles of splintered rubble or completely, totally gone, and almost another 1,000 were damaged to varying degrees.

"An estimated 2,000 other buildings were also destroyed, or rendered unfit for use, most notably St. John's Regional Medical Center, an inter-connected, twin-towered facility of seven and nine stories. Home Depot, Walmart, and many other businesses and buildings in an industrial park in the southern part of Range Line Road, the commercial area of Joplin, were completely destroyed, with concentrated loss of life. The heart of the city was effectively decimated. Without power and communications as night fell, Joplin, Missouri, descended, darkly and mutely, into tortured chaos."

Patty McCarty suddenly filled Scott's iPhone screen.

"Scott, we'll break for thirty minutes now. Please wait for us to reconnect. Thanks."

Scott leaned back, sighed, and tossed the iPhone on the couch next to Katy.

"Let's go for a walk and a little fresh air. There's a warm morning sun shining outside, and it's definitely not happy time in here."

When the connection resumed thirty minutes later, Alda came immediately to the point.

"Let's conclude with a brief synopsis. As we finish, please do not hesitate to comment if you have additional information you consider pertinent."

The Deputy Director referred to the top page of several held in her left hand.

"An extreme weather-related catastrophe requires immediate attention to, and care for, humanity impacted by the event. First, life-sustaining medical aid to injured survivors, then timely supply of items necessary for continued survival including, but not limited to, food, water, and shelter. Activation of the state national guard to maintain law and order, and finally, the process of identification and appropriate release of the deceased as a result of the calamity. As you have detailed at length, this was, in contrast to Katrina, accomplished in a most admirable, ongoing manner at all levels. Thank God for that."

She shuffled to the next page.

"The rebuilding process in the affected areas began with removal of debris. Over 800 truckloads of debris totaling almost 3 million cubic yards, roughly more than twice the amount removed from the twin towers after 9/11, were hauled from Joplin. The cost of the removal to various landfills in the area, accomplished in three months time, was approximately 100 million dollars. Between fifteen and twenty thousand trees were destroyed, left splintered and bark-stripped with only a few stubbed branches remaining. The Army Corps of Engineers, assigned by FEMA, has supervised the removal and disposal process that used a thirty-ton wood chipper to produce huge amounts of mulch.

"By autumn, the path of destruction through the residential areas was essentially cleared, leaving only driveways and foundations, one after another after another, and nothing else. Occasionally, a set of concrete porch steps, but no porch, no house, just a flattened, scraped lot with a smattering of green here and there. Home construction has started in earnest, but entire blocks are still filled with vacant lots. Walmart opened a new store at 15th Street and

Range Line Road in November. Home Depot re-opened in January, with an unusual addition of a storm shelter inside the new building. The rebuilding of Joplin is acknowledged to be several years before completion is accomplished."

Alda leaned back and pointed to the next page.

"Housing for the displaced was put into place as quickly as possible. This was FEMA in proper action, in coordination with a state Disaster Housing Task Force, and in timely response with needed resources. This was well done, not like Katrina at all."

She referred to red-circled paragraphs and read:

"Based on the review of the task force assessment of housing needs, a mobile home program was instituted in the fastest time possible. Three areas on Highway 171, directly south across from the Joplin airport, were prepared and designated Hope Haven, Hope Haven II, and the third named for Officer Jeff Taylor who was killed by lightning the day after the tornado in performance of duty.

"Placement of single-wide mobile home disaster occupancy units reached a maximum approaching 600 in the week of October, 2011. As available housing in the Joplin area increased, the FEMA occupancy level was reduced below 500 during February, and further departures from the temporary housing has continued. There has also continued, unfortunately, a residual population that has experienced a high incidence of crime-related events including domestic violence, robberies, prostitution, and drug dealing."

The Deputy Director rubbed her forehead.

"There is always demonstrated, inevitably, the very best, and, of course, the worst, in behavior when a catastrophe highlights human nature."

Alda Martin Starkey pushed back from her desk and addressed Scott Wasson with emphatic directness.

"This was the deadliest tornado since records were started sixty years ago. A staggering 2.8 billion dollars in damage. The level of storm intensity in this, and Katrina... When, not if, these extreme weather events happen more frequently, how will we be able to respond as well as we did in Joplin?"

"The million dollar question, Mrs. Starkey... actually the multiple billions of dollar question, is how we will respond should several weather-based catastrophes occur not in a series of events, but close to, or on top of one another. How would we possibly be able to provide the necessary services if several simultaneous demands were made on resources that are, obviously, not unlimited?" Scott reflected her reasoning.

"Katrina was *F U B A R*, the World War II adage used in <u>Saving Private Ryan</u>: *fouled up beyond all recognition.* The Joplin crisis, in contrast, was well handled. There was coordinated action between varied levels of government, combined with incredible response from across the nation. All in all, it was a demonstration of America at its best. But what if Katrina and Joplin were separated not by six years, but six days?"

§§§

Chapter Eleven

South-Central Nebraska 2012
"Miller's Trucking"

She heard his heavy footsteps trying to dislodge the caked-on dust from his boots as he trudged up the back porch steps and shuffled into the kitchen. The squeaky screen door slammed behind him. Laura avoided eye contact with Bob because she could not bear to see the weariness and despair in his eyes. He slid into his favorite, comfortable chair nestled near the fireplace and waited for Laura to join him with some sort of drink.

They had been married many years, and their myriad small routines afforded him immense comfort and pleasure. Even her quirky determination to surprise him continually with an array of alcoholic concoctions - some with actual names and some with her personal experiment number - made him smile.

She knew from their conversations throughout the week that tonight was not a "Fuzzy Navel" or "Harvey Wallbanger" night, so she brought over a chilled pitcher of Sangria with lush, fresh-sliced oranges - a real delicacy - filled their glasses, then quietly curled up to listen once again to his trucker tales.

Bob and Laura had both been born and raised in south-central Nebraska. They were high school sweethearts and went to the University of Nebraska together. Bob's family owned and ran the largest grain-hauling company in central Nebraska. His great-grandfather had established the business in 1912, using horse-drawn wagons to help farmers move grain from their fields to the grain elevators and railroad station. As agriculture expanded, so did the trucking company, hiring more drivers, and keeping the most advanced grain-hauling equipment as an integral part of the company.

From the beginning, the family understood the value to the company of having the children be well educated. Consequently, all of the children went to college. Knowing that Bob would probably take over the business, Laura studied Finance and Economics. Bob, on the other hand, chose to study his passion - Literature. He figured he would have plenty of time to learn the business later.

Their early years were frantically successful. Three adorably precious and precocious little guys came very quickly into their lives. Bob was so proud. Laura was worn out. The new additions required major remodeling of the older homestead. Laura insisted on serious square foot expansion, which their ever-expanding company could easily afford. Luxuries became necessities, and the once sleepy little farmhouse began to look more like a destination resort.

Still, Bob and Laura remained loyal to the ideals of the great-grandfather. They were generous with the company's resources among family members, and everyone went to college. Family members were seen as valuable assets to the company and, as such, were expected to develop and contribute their individual talents for the good of the whole.

The concept had worked successfully for over one hundred years, so it was difficult for Bob and Laura to recognize when or why it all began slowly but inexorably to come apart.

"Finding a Fit for Mandy"

Jake and his dad went round and round about college and fulfilling his family obligations.

"... but I HATE school!" Jake would thunder.

"You have no option," would always be his father's steady reply.

Finally, Laura stepped in one day and made a suggestion she would later come to question in the quiet of the night.

"Jake, you love working on the company trucks and machinery. Would you consider going to a technical school to learn whatever it is that you don't already know about keeping all the equipment running?"

Jake was so surprised by the simplicity of the solution that he laughed. "Now that I can do," he chuckled.

They soon found that one of the top tech schools in the country for big rig repair and mechanics was located in a small mountain town in North Carolina. Bob and Laura liked it because Bob's childhood friend C.B. lived less than eighty miles from the school. Family would still be close by. Jake liked it because it was half-way across the country from home. Parents would *Not* be close by.

With the beginning of the next session, Jake was gone. Laura worried that sending him so far away might not have been a good idea, but Bob reassured her that Jake would be just fine. Although he and his son seemed to be at odds most of the time, Bob respected and admired Jake in ways his son never knew. Bob was reticent about showing his affection openly, and, despite his literary bent, he lacked the words to express his feelings to his son.

Jake had an innate gift for understanding the workings of all sorts and sizes of equipment and machinery, and he never seemed to mind patiently explaining to the old-timers yet again how something should be repaired. He was kind and generous and had a strong character....and was he a bit rebellious? Sure. But, then, so was his dad in his younger years. No, Bob wasn't worried about Jake at all. The time away from family would be good for him.

When Jake returned two years later, he brought Mandy with him. She was terrified that "the family" would not accept her, despite Jake's repeated reassurances to the contrary. She was from a rural mountain community, and everything about her revealed her lack of sophistication.

Rough, tough, independent Jake adored her. He couldn't wait to get her home. He wasn't sure what role she would play in the business structure, but surely someone as wonderful as Mandy would do something wonderful for the company.

Mandy had all the best intentions and diligently endeavored to make herself useful with few positive results. Perhaps people were just used to having things done their way, not Mandy's. Perhaps some were just not comfortable with her Appalachian accent. Whatever the reasons, she continually struggled for acceptance.

Laura, once again, was the problem-solver. This time, however, Laura felt that her own sanity was also at stake. She had come to the realization that when the wives of her boys would come to work at the company office located, of course, in the compound, they would bring their babies and small children with them. Laura was being inundated by toddlers! And who loved playing with and caring for the little ones the most? Mandy.

To Laura, the solution to the dilemma was obvious. Build Mandy and the children their own playhouse and incorporate it into the rest of the property. Let Mandy help design it. Make it as whimsical and colorful and fantastical as they could imagine. Have lots of little water sprinklers coming out of lollipops and mushrooms and slides to help keep the children cool and a white picket fence to help keep them in. *She will need a cook to feed them and a maid to help with the messes,* Laura mentally added.

Mandy was in heaven, and the other parents knew that Mandy would love and care for their children as much as if they were her own.

§§§

Chapter Twelve

Video Conference, Miner, MO -
Washington, D. C.
Monday, July 5, 2012, 9:08 AM

Scott Wasson was somewhat surprised when Patty McCarty's image came up on his iPhone.

"Well, good morning, Mrs. McCarty. Everything all right?"

"Morning, Scott," Patty chirped. "No problem, just a busy start this Monday. Alda's been called out of the office, unexpectedly, for a round table meeting with visiting dignitaries. We weren't scheduled to participate, but someone asked for her to sit in, if possible. Carter's swamped administratively, and Justin is out on assignment. So, in place of the Deputy Director, here I am. How about you two, doing O.K.?"

"We're fine, thank you. We're parked at Lambert's Cafe in Miner, Missouri, at the moment. Katy fixed coffee and muffins earlier, since Lambert's doesn't open until 10:30 this morning. We'll go in for lunch at noon, Mrs. McCarty, but until then we're double-checking our Mississippi River plan."

Katy slipped the iPhone from Scott's hand and walked towards the kitchen.

"Hey, Patty! Hang on while I get some coffee."

Scott, with his hand still in mid-air, watched with mild fascination as his presence became irrelevant.

"Hi Katy! I can't wait to hear about your visit with Harvey and Grace. I've got some time, so tell me. I'm all ears."

"Well, it was good, but really sad. I haven't seen my grandfather and Grace since my parents' funeral, and that was the summer of my freshman year. A lot has happened with them in the last four years."

"You said Grace mentioned there have been ongoing health issues."

Katy sighed, "Yes. My father was Grandpa Harvey's only child. Soon after my parents were killed in the car accident, my grandfather had a heart attack, and he's had two more and several strokes since then. He's not in good shape at all."

Patty's voice softened. "I'm so sorry, Katy, but you were able to have a good visit with him?"

"Not like when I was little. My grandpa was always full of life, and always had a teasing sense of humor. That's pretty much all gone now. It was hard to get him to talk, and then he said but only a little before he lost attention. He rarely gets out of the house, and Grace says that's just as well, since it's been an extremely hot and dry summer. She said that when she stepped out the front door the day before we arrived, it was so hot that it took her breath away. It is hot here, no question."

"Well, we've been watching the weather in Missouri from here, and over one-hundred-degree days started about the time you and Scott arrived in Joplin. And all, I mean all, of Missouri is in some state of drought. Severe at the least, and a good part of the state is in extreme drought. And down around Poplar Bluff in the boot heel? That area is actually in exceptional drought, the worst rating there is."

"Don't we know it. Scott and I were going to take Grace and Grandpa in the motor home to watch the West Plains "Skyfire" Tuesday night at the airport, but they canceled the fireworks display because it was so dry. And south of here, at Thayer just above the Hardy Hills in Arkansas, their fireworks were canceled, too. So we didn't get to take them anywhere. The Fourth of July without fireworks just isn't the same."

"Katy, if you ever have the chance, be in D.C. for the Fourth of July. Guarantee, it'll be a life-long memory. There are parades and concerts all day long, and the most spectacular fireworks display you can imagine over the Washington Monument at dark, about nine o'clock. But research and plan in advance, because there's always a lot of people in town for the celebrations. And although you can certainly move around freely, keep in mind that security is tightened for the day. Anyway, it is too bad you weren't able to get Grace and Harvey out to see some fireworks."

"It is too bad, but I guess I can understand it. Grace said there hasn't been any real rain worth mentioning since the beginning of May, and wildfires have been way, way up. She said they saw on T.V. that there's been well over a 100 % increase in wildfires this year in the May to June months. And that doesn't count the numerous fires burning in the million-acre Mark Twain National Forest. Can you imagine, since the beginning of January, over 2,000 fires that burned over 25,000 acres? Fifteen homes and 150 outbuildings gone. Not out west, but in Missouri?"

"Carter said the other day that nationally, for the first half of this year the temperatures were the highest on record. So, it's a lot of places, not just where you are right now, if that's any consolation."

"Scott just pointed at his mouth, Patty. Does that mean he's hungry again? We just had breakfast a couple of hours ago. Maybe he thinks we've talked long enough. Now he nodded his head, "yes." Why do men think they know when we've talked enough?

"I'll leave you with this before I go. This morning, I was thinking it was good to be out away from the big cities, you know? But even in the southern reaches of the *Show-Me* state, we're still impacted by all this unusual weather. What's generated in the density of New York, or Chicago, or L.A. affects everybody, everything, everywhere. Our ever-expanding density - concrete is our soil, and oil is our lifeblood. And it goes on and on. I think we're the kudzu of the animal kingdom."

"That's funny," Patty laughed, then stopped. "Hmmm... Maybe it's not so funny."

"Bye Patty."

"Bye Katy."

§§§

Chapter Thirteen

Joplin 2012
"The Phone Call"

In August, C.B. and Vanessa went to Joplin for Jordan's college graduation and for her marriage to Kyle Beeman. Remnants of the tornado could still be seen: twisted signs, debarked trees, huge dirt movers, and massive open spaces where houses and businesses once stood. The skeleton of the tilting, nine-story St. John's Hospital still stood as a stark reminder of the massive power of the monster that had descended upon the unsuspecting community. The steel-and-concrete structure, larger than a city block, was reported to have been moved four inches off its foundation by the multiple-vortex behemoth.

Jordan and Kyle's church sanctuary remained under tarps and in the midst of reconstruction. Still, their wedding was joyfully celebrated by all of those who had so lovingly welcomed Jordan into their lives after her dad helped her flee Katrina as that mega-storm began its assault on New Orleans.

In retrospect, Jordan came to appreciate that she was one of the luckier survivors of one of the nation's worst natural catastrophes which took over eighteen hundred lives and left a proud, historic old landmark in ruins. Her father was the one who had fought his way into New Orleans, while thousands were fleeing, to pull her out of harm's way even when she failed to see the danger. He was the one who found a trailer and packed as many of her belongings as precious time allowed, and it was her dad who made the right choice to drive north instead of trying to flee on Interstate 10 heading west.

I-10 from New Orleans to Houston became a deadly parking lot as untold thousands of cars slowed, then stopped, with nowhere to go and no means of escape.

Once it leaves New Orleans, Interstate 10 is elevated, carrying vehicles over miles and miles of bayous, marshes and swamps with few areas of hard ground to drive on. Extensive studies have since been conducted to prevent bottlenecks such as those that developed on I-10 from occurring again, but the revised contingency plans came too late for those people who were trapped in their cars for days without food, water, or basic sanitation. Military helicoptors finally began dropping MREs (Meals-Ready-to-Eat) and water along the route, but for some, it was not enough. Heart attack, stroke, dehydration, heat, running out of medications. There was never an exact count of those who lost their lives trying to get away from Katrina on that stretch of road.

The northern route that C.B. chose took them over the twenty-three miles of the open waters of Lake Pontchartrain. The leading edges of Katrina were already over the the suburb of Metairie when the wind-driven sheets of rain forced a halt to their packing, and C.B. had adamantly said, "It's time to go." Given the conditions at that point, choosing the Causeway had been a gutsy decision.

The full realization of what they had escaped came ten days later when they returned to New Orleans to collect the belongings that they were forced to leave behind and to look once more for Sadie, Jordan's cat.

As father and daughter returned to the city from the west, they were stopped at a roadblock. Interstate 10 through the city was completely shut down as parts of it were still under water. C.B. was forced to turn around, head west for some miles, then find an old two-lane highway that would take them through some small towns and, eventually, to another roadblock and checkpoint.

There, they were carefully quizzed as to their purpose for entering the city. They needed to show identification and were told that whatever they needed for survival – food, water, flashlights, matches – they needed to take with them. They were also warned that if they were not out by dark, they should stay where they were. There would be no "911" services available to assist them. It was obvious to C.B. and Jordan that the phrases *Enter at your own risk!* and *You're on your own!* were the operative terms of the day.

They rode in stunned silence as C.B. slowly maneuvered their pickup through the miles and miles of carnage. It was not only the shards of glass, the twisted metal, the downed power lines. The lumber was splintered, then strewn like pick-up sticks thrown in the air, only to fall back to earth in a nightmarish heap, daring someone to try to move one stick without having the rest collapse.

It was the personal items that seemed to touch most deeply. Chairs, beds, an oddly standing door frame, clothing, a swing set somehow pinned in a pile of cars and splintered wood, computers, a child's crib. It was as if houses without roofs had been picked up, turned upside down, and shaken, scattering their contents carelessly as a child discards an unwanted toy – all remnants of shattered lives left behind by an indifferent combination of roaring, commanding wind and piercing, torrential sheets of rain and unrelenting heat.

C.B. was particularly amazed at the sight of an exceptionally large, steel-beamed gas station canopy. It had been bent to the ground and twisted into some sort of surreal art form. *What kind of power,* he thought to himself, *could possibly have done that?*

Man's hubris allows him false assumptions about his ability to direct, corral, harness nature's power. The ultimate truth is that nature at its zenith is beyond man's capacity to control.

When they reached Jordan's apartment, the door had been torn from its hinges, the windows had all been broken, and everything of value that Jordan had left behind was gone. Little remained that had not been damaged by the elements. It appeared that a person or persons had camped out there.

As Jordan wandered around, she tried to pick through her ravaged belongings and salvage what she could. It was impossible not to think of the terrifying possibilities she might have faced had she remained there alone. No food. No water. No electricity. No law and order. No way to defend herself. The thoughts were chilling.

Jordan never saw Brent again. When the city flooded, so did the jail. A decision had to be made quickly pertaining to the safety of the inmates. The final decision? Free them all! Where they went and what additional havoc they wreaked upon that city already devastated beyond the breaking point? Who knows? But that's another story for another day.

C.B. and Jordan looked for Sadie, but to no avail. He never expected to find her, but Jordan had to try. Finally, she tracked down the manager, turned in her key, and she and her dad again turned their truck toward Joplin and her new life.

As her dad wove their way once more through the chaos of the devastation, Jordan surveyed the vastness of the disaster and chuckled. The metaphor was too rich in a perverse sort of way.

C.B. raised his eyebrows, "What do you find humorous?" he quizzed.

"The metaphor."

"Which is?"

"The wreckage of my life. Look around. We're leaving it all behind."

He turned to her and, with a wan smile and a squeeze of her hand, he nodded his head. He understood.

By the time they reached their safe harbor, Jordan was physically and emotionally exhausted. She slept for several days. When she awakened, she found her old and new friends, as well as a new church family, were willing and able to help her tie up the loose ends of her past life and establish herself as a Joplin resident. It was time-consuming and sometimes painful, but the end result was exhilarating for her. She was ready to start again, or so she thought.

Sometimes it takes longer for those emotional, internal injuries to heal than one thinks it should. Such was the case for Jordan. This was her new life, after all. Why wasn't everything perfect right now? Patience had never been one of Jordan's stronger characteristics, and, usually, when she needed it the most was when she had the least of it.

She still desired the marriage, the white picket fence, the swing set in the back yard. Only this time she was working hard to do it right. She found a bookkeeping job and a decent place to live, and she went back to school. She had always been a gifted student and took pleasure in strong academic achievement.

It was at this point that Kyle entered the picture. He was handsome in a rugged sort of way, and taller than Jordan, which was always a plus for her 5'9" rather statuesque frame. He had a sardonic wit, wrote poetry, and, even at thirty-six plus, still preferred role-playing games with a few close friends to large, social parties. He was a computer whiz for the VA, apparently. His family had lived in the Joplin area all of their lives, and Jordan had met him through friends of hers from church. They seemed very much alike in myriad ways, and Jordan seemed truly happy. Beyond that, to C.B. and Vanessa, Kyle was an enigma.

Jordan's parents would like to have known much more about the history of this young man who professed to love their daughter to the point of making a commitment of marriage. Neither of the young lovers said much, even when questioned. Jordan said that it was Kyle's life, not hers, to share if he chose. Kyle was reserved and cryptic.

In the end, they reminded themselves that it was Jordan's life, not theirs. Besides, she had done everything they had asked of her. She had gone back to school and worked hard, making stellar grades and earning top honors and scholarships while working full-time. She had started going back to church and was associating with normal, every-day goofy people, rather than the Yahoos that almost destroyed her life in New Orleans.

C.B. and Vanessa recognized that it was time to let her be the adult she already was, so, with extraordinary pride in her accomplishments, and a bit of trepidation in their hearts, they gave the bride and groom the best wedding their retirement savings could afford. The exquisite bride brought tears to Daddy's eyes; the vows that the groom had written and pledged to his bride brought tears to Vanessa's.

"It was worth it," the parents of the bride reminded each other as they began turning the lights out late that night, and again as they tumbled into bed, exhausted. "It was worth it!"

Their plans for the day after the wedding were to repack their car and prepare for the next leg of their journey – a visit to Nebraska. They were to leave early the following morning.

They were both looking forward to seeing Bob and Laura again. Bob and C.B. had been childhood friends, keeping in touch for over sixty years. Bob and Vanessa became instant karaoke buddies, although Vanessa had to confess that Bob really had the better voice. As for Vanessa and Laura? Well, they could have been sisters, they were so nicely matched. Bob had even planned a big barbecue at the homestead when they arrived. Sadly, it was not to be.

The night before they were to leave, Vanessa received a call from her sister

Peggy. Carol, Peggy's eldest child, lay in critical condition in a Fort Worth hospital.

"Hurry," Peggy whispered through her tears. "The doctors don't think she's going to make it."

"But... what happened?" was all that Vanessa could get out.

"Car accident. I... I can't talk. Are you coming?"

"Yes, of course! We're in Joplin still, but we'll head your way first thing in the morning. We can stay at Nick's. Oh, Peg, I... can't believe this is happening.... I love you. See you tomorrow."

"I'm so glad you'll be here. The pain is unbearable. She kept telling me she was OK. ..I believed her... and now I may lose her." She sobbed again, then managed, "I love you, too. Bye."

With that, the phone went dead.

§§§

Chapter Fourteen

Video Conference, New Orleans -
Washington, D. C.

Monday, July 23, 2012, 9:01 AM

"State of the Heartland:

The Mississippi River Valley and

New Orleans, Louisiana"

The Deputy Director's voice was clear and upbeat as the video connection began.

"Good Morning, Mr. Wasson. A fine morning, actually. No early Monday morning crisis, and so far, no interruptions as I've read over your report. Well done, once again."

"Thank you, Mrs. Starkey. Katy does have a talent for formatting a great deal of information into a coherent package."

Alda leaned back, then asked, "So, before we begin, where are the rolling stones at the moment, and what are your traveling plans?"

"Well, after touring the areas that flooded around the French Quarter, we settled on Veteran's Boulevard at Transcontinental Drive in Metairie. We're in the parking lot of the family-owned Zuppardo's supermarket. Unusual name, but it's an excellent grocery store with a wide variety of goods, and it has a small, older-fashioned, short-order, sit-down eatery. Full breakfast, bacon and eggs and grits, plus sandwiches, basically. We've eaten there several times, and shopped the store to stock the motor home for the road.

"As soon as we've completed your review of the report, Katy and I are ready to roll. But before we leave, we're going to cross Vet's Boulevard at the first turnaround east to Gambino's Bakery. The locals say it's a *must stop* for a taste of traditional red velvet cake and the like.

"We'll probably head west on Interstate 10 to Lake Charles, then turn north on Interstate 49 toward Shreveport. We've been told there are some interesting bed-and-breakfasts in Natchitoches, just off Interstate 49, a town where Sam Houston is said to have stopped on his way from Texas to Washington, D.C. Just a small city now, it was once expected to be larger than Shreveport, until the Red River changed course, and isolated the town. That's where we plan to overnight. The next day we'll pick up Interstate 20 at Shreveport, and from there, it's all the way west straight into Dallas."

The Deputy Director nodded and reached for the report.

"Good. You've done very well, Scott, and I'm more than satisfied with the results of your efforts to date. We can only hope your

111

trip continues to be as smooth as it has been so far. By the way, I've heard that your daily debit card check-ins have usually been made by Katy. Evidently, Patty and Katy have developed an almost mother-daughter relationship, and I'm personally pleased that has happened as well. Now, to the report, and, of course, please comment if you feel it necessary.

"You've divided your report on the Mississippi River Valley into two sections: first, extremely high water levels in 2011, and second, extremely low water levels in 2012. Very heavy rainfall from tumultuous spring weather systems in April and May of 2011 resulted in abnormal amounts of water to run from the tributaries into the Mississippi where it borders Illinois, Kentucky, and Missouri.

"Ironically, the systems of levees along the Mississippi acted as a choke on downstream water flow and caused those tributaries to back-fill, then over-crest and flood. The levees on the west bank of the Mississippi in southern Illinois were actually blown open to save the city of Cairo, but caused over 125 thousand acres of farmland to be flooded in eastern Missouri.

"The river levels increased as additional waters poured in along the way from more tributaries feeding the Mississippi River as it meandered south between

Arkansas and Tennessee, and Louisiana and Mississippi to the Gulf.

"Vicksburg and Natchez in Mississippi set all-time records as the Mississippi River crested through those cities. The huge Morganza Spillway, above Baton Rouge, was partially opened for the second time in history, which flooded almost 5,000 square miles. Of the 350 bays in the Bonnet Carre' Spillway upstream from New Orleans, 330 were opened as the flood waters reached lower Louisiana. Parts of western Kentucky, Tennessee, and Mississippi were declared federal disaster areas, with many evacuations, numerous deaths, and estimated damages in the billions of dollars."

Mrs. Starkey took a deep breath, then continued.

"Then, in 2012, the water levels have flipped from extreme highs to extreme lows. The town that was saved from floods by blowing levees to drain high water levels in 2011, Cairo, Illinois, now, one year later, has historic low river levels.

"At Vicksburg, Mississippi, water levels have crested in one year a difference of an astounding fifty feet. Memphis, Tennessee, same type of change. Floods last year, levels so low this year that river commerce has been severely affected.

"Barges on the Mississippi River move everything from coal to grain valued at almost 200 billion dollars of commerce. One barge has the carrying capacity of sixty tractor-trailer rigs, yet with lower transportation costs. This year those barges are risking running aground. Convoys of barges are stranded in some areas, waiting for dredging operations to deepen channels for passage of commerce in both directions. The net result is a severe bottlenecking of trade on a transportation system more important than any other to the economy of the United States."

Alda Martin Starkey glanced up from the report with a grimace.

"And this, New Orleans, seven years after 80 % of the city was under water, and almost 2,000 residents died, there are still abandoned houses everywhere. As you point out, hurricanes rotate counter-clockwise, causing the greatest storm surge to hit the areas to the right of the eye of the storm as water is swept up from the sea beneath. All this rebuilding effort that you have documented was the result of destruction to the left, the weaker, side of Katrina.

"What will happen when another super-sized storm comes ashore with these fifteen-to-twenty-five foot storm surges? And

114

not just along the Gulf coast. The eastern seaboard, from Florida on up, is vulnerable as well. Think of Hurricane Hugo in 1989 that caused almost thirty deaths in South Carolina and over 10 billion dollars in damage. If a huge hurricane should sweep all the way up the east coast... "

She paused, then looked directly toward Scott.

"Curious, is it not, that the Mississippi River Valley should flood in 2011 to a record-setting crest in Vicksburg, and one year later Vicksburg river levels are at historic, near record-level lows? In May 2011, a record-setting crest of fifty-seven feet, and in July 2012, a low level of under two feet. Surely, that is an undeniable example of back-to-back extreme weather events. It's as if the climate is swinging back and forth, out of balance."

"Mrs. Starkey, I'm convinced, and have been for some time, that our climate is out of balance. I think of a finely-tuned internal combustion engine as an example. At the heart of the engine, a crankshaft, driving pistons up and down cylinders, in tune with timed combustion, combined with literally hundreds of moving parts. If all components operate as intended, it's a smooth-running machine, but if any essential part, or parts, perform out of specified parameters, the engine will lose its natural balance, vibrate, shake, lug, chug — any of a number of not good conditions. Undesirable, for those dependent on the ride of the vehicle driven by that engine. The entire population of humankind is on this ride of this planet, and we're tinkering a smooth-running machine into a state of imbalance."

A slow, sad smile came over Scott's image.

"You know, Katy was raised in a small north-central Illinois farm town. Granville, with a basically constant population of about one thousand people for the last fifty years. We were talking last night about her early years there, and what was in store for all the people she was raised with, as the weather will become more and more out of balance.

"After a quiet moment, she told me they had a top-loader washing machine, and while she and her mother were doing laundry one day, a heavy load of towels shifted to one side. Katy said she was frightened to tears as the washer began to vibrate in the spin cycle, then started to jump and bounce around with loud banging, before her mother quickly shut off the power to the machine."

The smile evaporated, and his usually-shaded impatience blazed brightly on Alda's screen.

"We won't be able to just turn the power off to stop the vibration when it begins in earnest, so we better start to get the balance back in our machine while we can. That means now, today, big time."

Scott tapped 'End' on his iPhone, and disappeared from the Deputy Director's view.

§§§

Chapter Fifteen

Wichita, Kansas 2012
"What's Wrong With Betsy?"

Stacy was thankful to see Mrs. Hadley finally walk out of her shop. It was another miserably hot day, and her feet hurt. So, of course, today was the day Mrs. Hadley wanted extra conditioning for her hair and a neck massage. Then, when she tried to pay with a credit card, it was declined.

"I just know I have money in that account," Mrs. Hadley protested weakly.

"It's probably a delay in money transfer transactions," Stacy responded absent-mindedly. "I've had it happen to me. Don't worry about it."

She accepted Mrs. Hadley's check and asked casually if she wanted the check held a few days. After a moment's hesitation, Mrs. Hadley declined, thanked Stacy for her gracious offer, and left. As Stacy locked the door, turned out the lights, emptied the cash drawer of what pittance there was in it, and grabbed her purse, she reflected once again upon how lucky she and Mike were still to be employed in Wichita, Kansas, in the summer of 2012.

Her little beauty shop had somehow managed to keep bumping along. She seldom had new customers, although occasionally girls from the college would drop in unexpectedly and be desperate for a haircut or a style for a party, and could she please work them in. She always did. Her "bread and butter" came from her regulars, which she liked, for the most part, and really appreciated.

The real family support came from Mike's job at Aero Jet, an airplane manufacturing company crafting exorbitantly expensive, private luxury aircraft for exorbitantly wealthy clients. His longevity and good work ethic had protected him from lay-offs over the last few years and had helped them stay afloat when others went under and had to move. All things considered, he and Stacy had a pretty decent life.

It's funny, Stacy mused. No matter how broke the country seems to be, there are always millionaires and company executives and corporations who can afford their own private planes. Stacy shook her head again at the thought of it, sighed, and turned her not-quite-new Chevy into their driveway.

There were "For Sale" and "Foreclosure" signs everywhere. What the recession hadn't broken, the drought seemed to be unhinging. Stacy had heard on the news that the drought had laid waste to almost 60% of this year's Kansas farmland. *Surely not*, as she thought about it later. *I just must have heard that wrong. I need to go back and check on that when I have a chance.*

Unfortunately, Stacy was one of those people who never went back and checked on things. To Stacy atmospheric imbalance and heat waves and drought merely meant another hot Kansas day with a bit of water rationing thrown in for good measure. All Stacy knew was that her feet hurt.

Mike met her at the door with eight-year-old Betsy in his arms. He spoke so quickly that Stacy hardly understood him.

"She's really sick. Temp is 104*. Vomiting. Diarrhea. Head hurts. The doctor's waiting... wants to see her right now... I'll call as soon as I know something."

With his listless daughter cradled on his shoulder, he was gone. Stacy was stunned. She seemed fine this morning, a little draggy perhaps, but nothing way out of the ordinary for a late night/early morning routine. Then she remembered that Betsy had been tired last night and had gone to bed early.

That wasn't normal for her always-ball-of-fire Betsy who couldn't wait to be the first one out the door to see what the world had to offer. *Oh, my God.* She slowly turned from the closed door and stared at nothing. *What is wrong with my baby?*

Almost two hours had passed before Mike called.

"We're at the hospital. They've put her in an isolation unit in ICU. They just aren't sure yet. It could be meningitis or encephalitis or some other sort of virus. Whatever it is, honey, they're pulling out all the stops."

They stayed by her bedside for two weeks as their courageous little girl fought off the West Nile virus and battled her way back to health, weaker, but alive. They could not understand how a tropical disease like West Nile virus could almost kill their precious Betsy. They did not live in the tropics, after all. They lived in Kansas!

§§§

Chapter Sixteen

Video Conference, Dallas, TX -
Washington, D. C.
Monday, August 6, 2012, 8:58 AM
"State of the Heartland: The West Nile Virus"

Carter and Patty greeted Scott as the video connection focused and cleared on his iPhone.

Patty pointed to the folder on the Deputy Directors' desk, "Your report is in, but Alda's not. Today is her father's sixty-seventh birthday, plus Admiral Starkey will arrive several days from now for an overdue shore leave. Can you believe being born on the very day regarded as the dawn of the atomic age?"

Carter's smile dropped with Patty's question. "One hundred thousand city residents on one side of the balance, and one million human beings on the other side, with one man at the pivot of decision. Sixty-seven years ago, Hiroshima, chosen as the head of the pin for the atomic angel to sit on. Just about forty-five minutes ago, their time. Humanity does put itself in unfortunate straits on occasion."

"Continually," agreed Scott. "With us, the greater the promise of our rise, the greater the danger of our decline."

"Well, now that you two have brightened the beginning of the day, where's Katy?"

"Indisposed, momentarily. So, did Mrs. Starkey review the dispatch over the weekend, leave any comments or questions?"

"Actually, no, Mrs. Starkey did not, Scott. However, she did leave instructions for us to do so in her stead. Both Patty and I have finished our read, but Justin isn't in yet. Would you care to wait for his input, and have us contact you at that point?"

"Thanks, Carter, but I don't think so. If Justin has questions or comments, he can contact us while we're on the road. It's really hot here, and we want to get moving north, though I think we'd have to go a long, long way up toward Canada to get cool. It definitely would be nice to be in the Northwest again, back home in Marysville, Washington, about now. It's been at least 100 degrees every day but one out of the ten or eleven days we've been in Dallas, and several days it reached 107 degrees. This is too much for me.

"At any rate, much of this report, as you've seen, is mostly data-driven, incidences of infection and deaths. We've had little need for much interaction with the locals here."

Carter sat back, fingers intertwined at his chin and asked, "I have one question. Do you see a correlation between the spread of West Nile virus in the heartland and the effect of global warming and atmospheric imbalance?"

"Frankly, that's difficult to ascertain, Carter, but my educated guess is *no*. WNV was introduced to the U. S. through New York City in 1999. That fact has been pretty well established. But since then, WNV has spread across the country east to west, north to south, concentrated mostly in dense urban areas, here in the Dallas-Fort Worth metroplex, Oklahoma City, Wichita, Kansas. There have been many documented WNV infections, with only a very small percentage resulting in serious illness, and even fewer deaths.

"This year does appear to be headed for the record of most infections in the United States. So far, over one thousand infections and forty deaths have already been reported. Roughly 75% of those reported infections have occurred in Mississippi, Louisiana, Texas, Oklahoma, and interestingly, South Dakota.

"The high numbers in Texas have been especially unusual. Approximately half of the infections, 500 of the 1,000 reported cases, and again, half of the deaths, twenty out of forty nationwide, have occurred in the state, centered mostly here in the metroplex. It has been particularly scary for Dallas, with the result that repeated insecticide spraying has been conducted for the first time in almost fifty years.

"Anyway, we've checked out from the Hilton here at DFW Airport, returned our rental car, and have been shuttled out to long-term parking to the motor home. There was no way we were going to maneuver that forty feet through downtown Dallas, and we didn't know any safer place to park it than at the Airport.

"The problem is, if you're not familiar with the metroplex, it's impossible to determine one area from another. The landscape is the same in all directions as far as you can see. The myriad criss-crossing freeways run past clusters of chain stores, Walmart, Staples, Home Depot, and the like, then pass through sprawling residential areas, only to repeat the cookie-cuttered sequence over and over again. Boggles the mind, actually. Katy hates it, and I can't say I'm very favorably impressed, either. So, we're really ready to head for Oklahoma City."

Patty outright laughed, "You're antsy, all right. Say 'hello' to Katy. Goodbye, Scott."

Carter smiled, waved a goodbye, then leaned forward and broke the connection.

§§§

Chapter Seventeen

Arlington, Texas 2012
"The Parents Visit"

By the time C.B. and Vanessa arrived in Fort Worth, Carol was gone. How could one who was so vibrant and enthralled with life no longer be a part of it? She lived on an inlet right at the point where the Chesapeake Bay meets the Atlantic Ocean, and when she was not lovingly nursing her elderly patients, she was on the water in her sailboat.

Apparently, the accident had actually occurred weeks earlier. She had allowed her wounds to be cleaned and stitched but felt that hospitalization was just a waste of money. She was, after all, a nurse, and she could tell if something were seriously wrong, or so she thought.

With the insurance money she received for the remnants of her vehicle, Carol bought herself another car and decided to drive to Texas to see family. She was surprised that she had to stop for sleep twice because her headaches were so severe, but she did not turn back. It seemed terribly important to her that she make the trip. She called Texas several times along the way, and those who spoke with her said that she just sounded tired.

She was able to reach her sister Pam's house in Arlington, Texas, before she collapsed and was rushed to the hospital where she lapsed into a coma almost immediately. The doctors cited massive head trauma and other significant internal injuries as they fought gallantly to save her life, but within thirty-six hours, she had slipped away. She was forty-nine.

The overwhelming shock and sorrow that Vanessa and C.B. felt over the sudden loss of their niece was assuaged somewhat by the true pleasure they experienced at seeing Nick, Kara, and their precious Ashley again so unexpectedly soon. They had been to Texas for a visit only six months earlier, but in the world of a toddler, that is a lifetime. After initial reservation, Ashley took them in stride and, finally, began calling them *Papa* and *Grammy*. C.B. and Vanessa could not have been more delighted.

Their time in Texas seemed to evaporate. Before the service there were myriad details that needed attention. Contacts to be made, a large dinner gathering of relatives and friends who had traveled from Virginia, the Carolinas (both North and South), Colorado, Arizona, and on and on. People wanted and needed time to arrive. Peggy vacillated between needing Vanessa around and just wanting to be alone. By the time the service had been conducted and most of the guests had left, C.B. and Vanessa had been camped out at Nick and Kara's for almost a week.

Ben Franklin's aphorism that appeared in Poor Richard's Almanac suggested that three days was long enough for any house guest. He was, in fact, more direct. The quote actually read, "Fish and visitors smell in three days." Even if Nick and Kara remembered the quote, they were gracious enough not to remind their parents of it, and not once did they suggest a move to a hotel. Solitary people though they were, they seemed to enjoy having family around for a change.

One of their evening conversations turned to the topic of the weather. It was August in the metroplex, and the temperature had passed 100 degrees for more than a month with more hot weather on the way. The West Nile virus had reared its ugly head and had taken twenty-four lives in Dallas County while hundreds more had become seriously ill with the disease. Dallas County even ordered aerial sprayings of insecticides for several nights in a row, an action that had not been taken in forty-five years, when an encephalitis outbreak took the lives of fourteen people.

Kara, normally a hard-to-ruffle-feathers mom, had some difficulty hiding her concern for Ashley's outdoor activities. Several of the Nile deaths had been quite close to their neighborhood, which meant that the mosquitoes carrying the virus were nearby, too.

Kara was not comfortable discussing any of the climate change-related issues. She was an exceptionally bright young woman who not only understood all the figures, the graphs, the charts, the statistics, but she perceived quite clearly their implications for her world and her little family. Still, she sat quietly and listened as the others shared what they had learned.

Nick began by debriefing them all on his latest environmentalist colleagues' meeting. One man had brought a friend of his from the GIS office in Houston. The friend was actually one of two environmental scientists in that office. Their weather concerns focused on rising sea levels and how they were impacting the shipping channels and the low lying areas in and around Houston. The scientist pointed out that because the polar ice sheets were melting so rapidly, discussions were currently underway about which areas would need to be protected and which areas must be allowed to sink into the Gulf.

"Really?" Kara interjected. "They're just going to let the ocean take people's homes and businesses?"

"They won't be able to stop it, when the time comes," C.B. responded. "Besides, I'm not surprised at all. I would be more surprised if other coastal cities aren't already having those same discussions. A recent video we watched showed what a sea-level rise of as little as one or two feet would do to New York, New Orleans, and a good portion of Florida...to say nothing of the rest of the world. "

"I think it may have been that National Geographic video <u>Six Degrees Could Change the World</u>. It was based on Mark Lynas's book of the same name," Vanessa spoke up. "I'll tell you what scares the daylights out of me. Lynas is projecting the Earth's tipping point could be reached as early as 2015 - that's not even three whole years away."

Kara interrupted again, this time with some urgency in her voice, "Now wait a minute. My understanding of the term *tipping point* is that it means the point at which humankind will no longer be able to stop the effects of the Earth's atmospheric imbalance, right?"

While she was talking, a tired little Ashley crawled into Kara's lap and lay her head upon her mother's shoulder.

"Yes, simplistically, that about sums it up. At that point, the planet will be like a runaway locomotive, the result ultimately....," she paused, looked down.

"NO!" Kara cried with tears in her eyes. "That's not possible! It just can't be."

At that moment, she pulled Ashley closer to her.

Vanessa moved to the couch, sat down, and swept a whisp of hair back from Kara's face.

"Oh, my dearest Kara. Scientists have been trying to warn us that this was coming for years, but there were big people with big power and big money who did not want the public to believe that the danger was real. Well, sweetheart, it is real, and there is very little time left to change the course of events. But, you know what? I am an optimist, and I believe in the resilience and imagination of the human mind.

"Surely, some scientist or engineer or college kid will figure out how to get all the harmful stuff out of the atmosphere before it's too late. Until then, it's our duty to survive... and to help the Earth in every little way we can. Now dry your tears, darlin', we all need to be strong for each other."

Kara leaned over and kissed Vanessa on the cheek.

"I'm going to go put this precious little bundle to bed. She's getting heavy," and she rose and left the room.

§§§

Chapter Eighteen

Video Conference, Hastings, NE - Washington, D. C.
Monday, August 20, 2012, 9:00 AM
"State of the Heartland: Heat and Drought"

Scott Wasson stiffened slightly with surprise as the Face Time image on his iPhone came into view. Seated casually erect to the left of the Deputy Director was a man unknown to Scott. *Alda has unquestionable presence,* flitted through his mind, *but so does this man.* Whoever he was, he had a strong-featured countenance, and his look was directed straight into Scott's eyes.

Alda Martin Starkey looked up from the red folder and noticed Scott's focus to her left. A glance at the man to her left revealed the mutually impassive gaze shared between the two men. She smiled knowingly at the young man's dispassionate display on her video monitor. *Ah,* she thought, *the initial canine-style nose-sniff.*

"Please relax, Scott," Alda soothed, "this is the Right Honorable, quite-recently-exalted, United States Naval Service Rear Admiral, lower half, William Henderson Starkey, my spouse. And, my dear husband, allow me to formally introduce Mr. Scott Wakefield Wasson, our investigative field officer-at-large, present to discuss his latest dispatch that will complete the <u>Report on the Heartland</u> series. Are we all now sufficiently at ease?"

"Admiral Starkey," Scott breathed in visible relief, and then recovered admirably. "A singular pleasure, sir, and one I've been looking forward to for some time. I did imagine a somewhat more cordially-inclined occasion, but be that as it may. Perhaps we will have an opportunity to share some time together when this project ends."

The Admiral flashed an engaging smile to Scott, "I would imagine that we will, son, and from Mrs. Starkey's comments concerning you, I expect to learn a bit more about a subject that has increasingly concerned me as time has passed. Now, however, I will leave you both to the task at hand."

The Deputy Director quickly re-entered the conversation.

"The Admiral would very much like to stay and listen to our discussion, I am certain. If you have no objection, I will allow him to remain as a silent observer."

Scott nodded assent.

"If the Admiral wishes to sit in with us, I am for it. And, as far as I am concerned, as an active participant as well."

Alda spread the contents of the red folder in front of her.

"Then that's settled. Stay seated, Admiral."

With a page of the report held aloft, she continued.

"Let's begin with the terms *imbalanced atmosphere* and *atmospheric imbalance*, which I have assumed to be one and the same. Those terms have repeatedly appeared, not only in this report, but throughout the entire series. You have stated that *imbalanced atmosphere* is the baseline causation for the entire sequence of severe weather events, which becomes, over longer terms of time, referred to as climate. For the sake of clarification, by *imbalanced atmosphere,* is it correct that you mean the change in the percentages, or ratios, of the chemical compounds that constitute the gaseous envelope around our planet?"

After a pause, Scott began his explanation.

"Think of it this way. For countless thousands of centuries, the basic composition of the air around and above the surface of the earth has been stable, essentially unchanging, and that has provided an element of balance to the greater, overall natural systems of our planet. Within a mere two hundred years, that stable condition has changed radically. The change matches, quite closely, the dawn of the industrial age, and the burning of fossil fuels to provide the energy to drive industrial growth. So, yes, simply stated, we have an imbalanced atmosphere.

"Furthermore, Mrs. Starkey, the greater the imbalance, the greater the change of the behavioral properties in that layer of gases around our planet. As the behavior of our atmosphere changes, other conditions in the operational systems of earth react with changes of their own. It is a reinforcing spiral of events in a vast system, incredible in intricacy, complex beyond comprehension. To make the situation even more interesting, it is an established fact that our current atmospheric imbalance is, by the way, increasing at an explosive rate."

The Deputy Director nodded.

"Thank you, Scott. Nicely stated. You go on to say that both the heat wave of this summer and the ongoing drought are in the category of extreme weather events. To be in that classification, they must meet the criteria of being no more than one out of twenty such types of recorded weather events in terms of intensity, duration, and so forth. Would you detail the reasons you believe these weather events, and the many others we have experienced as of late, have become so extreme, so unusual?"

Scott Wasson's gaze lowered for a moment, then regained directness.

"Basically, all weather events begin with the interaction of atmospheric conditions. In the polar regions of our planet, the sun's rays strike the earth at relatively flat angles, and a smaller amount of heat is received in those areas. The air above these polar areas is, therefore, cooled by the colder, usually frozen, surfaces beneath, and cold air masses are created.

"In the equatorial regions, the sun's rays strike the planet at relatively direct angles, and a larger amount of heat is received in those areas. The air above these equatorial regions is, in turn, warmed by the hotter surfaces underneath, and warm air masses are produced.

"The interactions which result when the cold and warm air masses coincide are very complicated, but we refer to the irregular lines on weather maps where they meet simply as weather fronts."

Scott stopped for several seconds, then started again.

"Now, please stay with me, as all this becomes a little bit more complicated.

"Our weather in the northern hemisphere is directly affected by the Arctic Oscillation, a periodically-changing condition originating in the polar region. The Arctic Oscillation drives cold air south, and how far down the cold air reaches and meets warm air determines the weather in that area.

"It is an acknowledged fact that sea water temperatures are on the rise, and as tropical storms develop into hurricanes, they have greater strength, power, intensity, and coverage area from the additional energy gathered from the warmer water.

"Abnormal behavior on the part of the Arctic Oscillation has also been observed for some time, which in turn, also causes unusual weather. When such divergent conditions combine . . . well, the result is a witch's brew we experience as increased extreme weather events: EF-5 tornadoes, severe droughts, and for another example, this summer's super heat wave."

Alda referred to time frames and temperatures underlined in the report, and read from them.

"The heat wave started in the last week or so in June, blazed unabated through July, and has continued now toward the end of August. Your report documents all-time record after all-time record exceeded in the southwestern states from Colorado to Texas, across the plains and Midwest, Kansas, Missouri, Arkansas, Tennessee, Kentucky, Indiana, Illinois, even Wisconsin and Michigan, and east to Connecticut, Maryland, New Jersey, D. C. , Virginia, North Carolina, and down to Georgia. So, most states east of the Rocky Mountains. New all-time high temperature records were set in Galveston, Indianapolis, St. Louis, Little Rock, Chattanooga, Nashville, Paducah, here in Washington, and Atlanta. In all my fifty years, I've not seen a heat wave of this intensity, this duration, this magnitude."

Scott agreed.

"Yes,Mrs. Starkey, it is very exceptional. Unfortunately, as the reports all summer have demonstrated, this is not an isolated extreme weather event. It's only one of the many dots that are related, that we need to converge into an over-all perspective. Just like the small, localized storm in Iowa, affected by the heat wave conditions, that developed into the destructive straight-line derecho at the end of June. Eighty-to-ninety mile-an-hour winds that swept east for over 600 miles across Illinois, Indiana, Ohio, Pennsylvania, and, finally, into Virginia. That was an extreme weather event in itself, resulting in twenty-two deaths and millions of dollars in damages."

The Deputy Director asked in exasperation,

"Where does all this stop, Scott? As you point out in the second part of your report, the drought is part of an over-arching, ominous pattern. The drought that has covered 80% of the United States in at least an abnormally dry condition is expected to last into November, and obviously will have quite an impact on crop production and consumer prices for food. This dryness is surely related to the heat wave we've just discussed, is it not? How do you account for all these extreme weather events that you have documented all summer?"

Before Scott could answer, the Admiral leaned forward and joined the conversation.

"I'm very interested in that explanation as well, but before you begin, I have a rather personal question. I would like to know how you came to be so attracted, infatuated if you will, to this area of science. Not your academic accomplishments, but from the very beginning, if you don't mind."

Scott Wasson sighed, and massaged his cheek before he started to speak.

"When I was in the fourth grade, nine or ten years old, I was introduced to General Science. I started thinking for long hours about all this that has become my life's work.

"Over time, I began to visualize myself above my house, then up in the sky until I was like a stationary satellite, in a fixed position to see my area in the United States, in the northern hemisphere, hovering over the Earth. From there I watched the planet, tilted twenty-three degrees off center, spin in a rotation from sunrise to sunset and back to sunrise, one complete revolution every twenty-four hours.

"I realized that with each daily revolution, the Earth moved 1/365th of its orbit around our sun. I understood that each day of spin caused a slightly different amount of sunlight to land on the same area of the planet's surface because the Earth was tilted that twenty-three degrees off center. As I put it all in motion, I grasped the concept of change of season. Spring, summer, fall, winter, one full cycle of seasons happening for each full orbit around the sun. One full orbit was one year, one full cycle. If I add more years, there were more cycles, and as they connected over time, the cycles became a frequency.

"In my mind, that frequency became the sound of nature, a steady hum, perfect in pitch. It became the audio background to the world of nature in motion, a video of the Earth moving through space in its orbit, trailing an unending equation of complex variables set in numerical arrays that described our natural systems, only a very few of which I recognized.

"With the exception of random events, such as the effect of an impacting asteroid, or the detonation of several thousand nuclear bombs, there is only one variable in that equation responsible for the change of the perfect pitch, that introduced a discordant whine, degenerated the beauty of nature around us, caused the deviation from optimum operational balance of our planet."

Scott hesitated for a moment, then continued.

"My dad had an old, yellowed cartoon clipping from the late 1960's taped to the roll-top desk in his study. The cartoon character was Pogo, and the caption read: *'We have met the enemy, and he is us.'*

"Simply stated, Admiral and Mrs. Starkey, if you want to know how I account for all of this abnormality, I'd say Pogo's statement was spot-on," and his image disappeared from the Deputy Director's screen.

§§§

Chapter Nineteen

Wichita, Kansas 2012
"What About the Farm?"

Stacy arrived home a few weeks after Betsy's release from the hospital to find the house quiet and Mike sitting in his chair without the TV turned to one sports channel or another. It was quite unusual for Mike because there were always big sports stories afoot...baseball trades, an NFL referee lock-out, bad calls by back-up refs, even the President had chimed in. "Get the veteran refs back in the game!" A chill ran down her spine that made Stacy catch her breath. Why was the house so quiet? *Do not panic,* she told herself. *Just take a deep breath, and find out what is the matter.*

Stacy put her stuff down, grabbed a Dr. Pepper, and sat on the couch. "So, what's going on?" she asked, trying to act casual in the process, a skill she had never quite mastered.

"Well, it's kind of a long story," Mike began. "Here, before I start, put your feet up. They're hurtin' tonight, aren't they? I can tell."

"Yea," she nodded her head as he scooted an ottoman under her feet. "Where are the kids?"

"They're out in back. I stopped at McDonald's after I picked them up so you wouldn't have to cook tonight. They'll be OK for awhile."

"Thanks, Mike. It's those little things you do that make me appreciate you even more!"

"Well, I don't know how much you're going to appreciate me when you hear everything that's going on."

"Oh, boy! This doesn't sound good. Well, have at it, Sweetie. You can't make my feet hurt any more than they already do."

Mike sat down and paused, not knowing exactly where to begin. Finally, he figured that he first started hearing rumors about big changes coming some months ago "around the water cooler."

"Nothing specific," he shrugged, "just tidbits. Upper echelon was not happy in Wichita. Might need to do something about it. Too many problems. Didn't like the weather. DIDN'T LIKE THE WEATHER? People howled when they heard that one! Heat? Drought? Wind?

"This is Wichita, Kansas, for Pete's sake. If the weather was a problem for you, why in the world did you locate your facilities in Wichita, Kansas?"

They both laughed.

"Anyway, the CEO ordered a climate overview and computer-model-based projections of climate patterns in the central plains. Apparently, he and the Board of Directors believe that the weather currently being experienced is not the product of a typical Kansas weather cycle, but rather a 'permanent weather change resulting from the fundamental rise in the normal temperature of the planet.' "

Mike stopped talking for a minute and lit a cigarette. He hated discussing this kind of stuff, because he thought it was just smoke-and-mirrors. He wasn't sure what the tree-huggers were going to get from making so much trouble, but he didn't appreciate the aggravation.

"Then the supervisors and accountants started putting together charts and graphs and statistics. Production down in the hottest months for four years in a row, heat-related injuries from hot tools or tools slipping from sweaty hands, heat-related hospitalizations such as heat stroke, heat exhaustion, and dehydration. One man passed out from the heat and fell from an hydraulic lift. Of course, he didn't have his safety belt on. Didn't think he was high enough to need it. He's still in a coma. Had he been any higher, he would have died immediately. Apparently, the insurance premiums are going through the roof, to say nothing of the water bill.

"The bottom line is that the executives no longer believe that Wichita is a 'viable' community from which they can run a successful manufacturing company. They are already making arrangements to move the necessary equipment and executives to Seattle."

With that, he rose and went to the kitchen. "Would you share some wine with me, baby?" Although he could have slugged down about three six-packs of beer right then, he knew beer wasn't her drink.

"That sounds really good about now. Where does all this leave us?"

"Well, there's good news, and there's bad news. The bad news is that I was not invited to move to Seattle. However, I was offered the opportunity to submit an application that would be carried to Seattle with Human Resources. They said it would be considered along with those of other applicants once they start hiring again. They intend to draw from the huge pool of Boeing employees who were not moved to South Carolina when Boeing made its move south. Who knows? I might look better than anything Boeing left behind.

"The good news is that I was given a three-month severance package that includes health insurance. Actually, this is really generous. Most employers don't give hourly employees much, if any, severance package at all. I, for one, really appreciate it! At least it buys us a little time."

Mike could see the tears begin to well up in Stacy's eyes and her nose start to get red. He always had thought Stacy to be a little cutie, but crying was never one of her finer moments. Red eyes, runny nose, blotchy neck, smeared mascara. No, he would rather see her smile. He quickly moved to her side, put his arms around her, and whispered, "We're gonna be alright, honey. You'll see. It's gonna be OK." Then he gently rocked her until his warmth renewed her faith and trust in their future.

In the days and weeks and months that followed, they continued to believe that something would work out for them. Mike argued repeatedly that he should go to South Carolina to see what Boeing might be offering as they set up their new base of operations. He knew the huge corporation had not moved thousands of employees all the way across the country. Maybe it was a long shot, but he needed to try something. Wichita had a plethora of aviation mechanics and no openings, nor could he see any reason to expect much in the way of future possibilities.

His severance package had included Career Counseling which suggested that he attend the nearby Wichita Area Technical College. There, he could refocus his skills and learn a new trade. He was still young, and it could be a great opportunity for him. What he recognized was that in the year...or two... of time it took to retrain, he would lose everything - his severance package, his insurance, his savings, his home, his way of life, his dignity. Stacy's little shop just could not carry them while he "retooled."

Late one night, long after Stacy and the children had gone to bed, Mike sat on his couch and closed the folder bulging with job-search pointers, the WATC app, and other well-meaning pieces of information. He leaned back, laced his hands behind his head, stared at the ceiling and began mulling over his options: see about what Boeing might have going in South Carolina, check out the tech training, keep applying for any mechanics jobs that come up.

There is always Stacy's parents' farm, he reminded himself. *They've offered us a place to live - plenty of room. They're getting older, and with the drought and all, they've fallen on hard times themselves and really could use the help.* He shook his head. *I just don't know what we should do.*

§§§

Chapter Twenty

South-Central Nebraska 2012
"The Playhouse"

By the end of the summer, all of the bright colors on and around Mandy's fairytale cottage had been sun-bleached or sand-blasted to barely recognizable shades of neutral. Most of the wonderful sizes and shapes and colors of plants and flowers that she and the children had so painstakingly selected and lovingly planted had died, and most of the grassy areas where the children played under the sprinklers was more mud than grass.

The men in the compound had gone to Herculean efforts to cover the children's play areas with pergolas and small gazebos and canopies and anything else they could come up with to shelter the kids from the sun while allowing the breezes to blow across the land. The shelters protected them from the sun's rays, but the air was still hot and dry and sucked the moisture from everywhere, including the children.

Mandy would set the tinkling little bells that signaled the end of outdoor play on timers so that they were never outside for too long. When all the small bells began to chime, the kids stampeded to the deck, which was also an outdoor shower area where the mud and dust were washed off them. Then, they were bundled in thick terry towels and escorted to their dressing rooms where fresh clothes awaited.

When Laura recognized that Mandy would need assistance, she considered qualifications much more carefully than she did expense. She had no idea when she selected Alana, a vibrant, experienced registered nurse, and Hannah, a gentle, thorough clinical dietitian, how vitally important her choices would become to everyone involved with Miller's Trucking. Her thoughts focused, of course, on the health and safety of all of their children, from the babies and toddlers all the way up through the young teens who also enjoyed going into the playhouse where they could play hide-and-seek or pirates with the smaller kids or just curl up in one of the quiet cubbies and read.

The inside of the playhouse was a wonderland for all of the children. Mandy's imagination had created curved walls and low ceilings and secret passages to hideaway corners. The lighting was built into every nook and cranny so no child would find himself or herself in a small cove that was suddenly verrry dark and scary. There was a child-sized pirate ship and twisty slides, and their napping beds looked as though they had been scooped out of the walls. Some were higher, requiring little ladders to access.

Others were closer to the floor so the smaller ones could crawl in unattended. They each, of course, could pick their own, and there were a plethora of choices.

Lunches were always scrumptious. It quickly became known that Hannah was something of a gourmet and enjoyed combining good nutrition with terrific cuisine. First, the teens were dropping in for lunch, then the moms wanted to eat with their little ones. Soon, the dads were scheduling their truck runs around lunchtime.

Finally, Laura had to step in. She was thrilled that Hannah's good cooking was promoting family togetherness, but Hannah was swamped with the amount of food she was having to prepare daily. The cottage had never been designed to prepare food for or to seat that many people.

A rotation system was established to limit the number of parents who could visit each day, and all teens over thirteen had to eat at the main house with the adults except on their "family day" at the cottage. How was anyone to know that Hannah's good cooking would lead to such a brohaha?

As the dry, searing summer heat continued to scorch the Nebraska plains, Mandy, Alana, and Hannah became increasingly concerned about the weather's impact on the health of their charges. Alana noticed that some of the children were no longer sweating as they played outside which was indicative of dehydration and possible heat exhaustion, both conditions that could become deadly quickly in small children.

The cool showers and electrolyte-fortified liquids with which they plied the boys and girls helped, but it seemed that there was always someone who needed cool washcloths applied. Mandy and Hannah made sure the children consumed their liquids while Alana checked temperatures and pulses. Then the boys and girls were put down for naps.

Late in the summer, Jake walked into the cottage to find Mandy sitting at one of the small tables alone and quietly crying. In two quick strides, he was kneeling by her side asking, "Mandy, darlin', what's wrong? Are you hurt? Has someone said something to upset you? What's goin' on?"

She picked up a napkin from the table to wipe her nose, then turned to him and cried, "I'm pregnant," and the tears began to roll down her cheeks again.

Jake was beside himself with excitement. "Really? Are you sure?" She nodded as she blew her nose.

"This is awesome! I can't wait! How soon? Wait a minute," he stopped.

He straightened up and became serious, "Mandy, honey... look at me. Why are you crying? What's wrong?"

She threw her arms around his neck and cried into his shoulder. He gently sat her back on the chair and asked again, "Sweetie, tell me. Why are you crying?"

"'cause I'm scared," she managed to mumble.

"Scared? About what?"

"About the weather and how bad it's getting and what if it hurts our baby?"

"Oh, my wonderful love," he whispered as he kissed her softly on a teary cheek. "Come on, let's go sit in the cushy chair."

He led her over to the huge stuffed rocker that could accommodate a reader and several small children. There, they curled up together, and Jake began to talk.

"You, my precious Mandy, are a part of me and the mother of my children. I will never let anything hurt you or my babies if it is within my power to keep you safe, and believe me, I will move Heaven and Earth to make it within my power.

"Dad and I have already been having serious conversations about how the changing weather patterns are impacting our way of life. It isn't easy for him to watch all his years of hard work, as well as that of his family for generations past, start to unravel. Each week he and my brothers are driving further and further out and are coming back with nothing but dark stories about no crops and small towns that are slowly but inexorably drying up along with the soil."

Then he told Mandy something she didn't know. "Alana has had several conversations with Mom about the danger of the heat, especially as it impacts the smaller kids. So Dad is taking that information into account, too, as he decides what our little tribe is going to do."

"I don't understand," Mandy interjected quizzically. "How does he think he can solve the weather problems?"

"Well, he can't. Not really, but he can make some decisions about how we're going to deal with it. He hasn't given me a clue about what he has in mind. All he said was that he's headed to North Carolina to talk to C.B. He really trusts his friend. Thinks he's a rational thinker and a good sounding board. You remember C.B. and Vanessa from Statesville? We met them a couple of times while I was in school.

"Anyway, that's where he and Mom are headed. Not that C.B. will tell him what to do. He has more sense than that. But Dad says they've always been able to brainstorm well together, and right now, Dad thinks he could use that from a friend. "

"But, Jake, what does all of that have to do with us and our baby?" she again quizzed.

"I'm not exactly sure yet," he responded as he wrapped his arms more tightly around her tiny frame. "But whatever happens, I am your man and your protector... yours and that baby's."

§§§

Chapter Twenty-One

South-Central Nebraska 2012
"The Unexpected Conversation"

After handing Bob his Sangria, Laura kicked her shoes off and curled up on the nearby couch to listen when he was ready to talk. She had long ago learned not to push him into conversation after he had been gone for several days. He would open up when he was ready. This time on his return, however, he was even more somber than usual. Part of their business included watching grain prices, weather reports, and soil conditions. They needed to know when the farmers were going to need trucks in their fields to haul grain to market. So, Bob and Laura were under no illusions about how bad the growing season had been for their customers. Still, seeing the withered crops and the pain in the farmers' eyes somehow made the reality even worse.

"The Sangria's good tonight, hon," Bob uttered quietly. "I really needed something cold to clear the dust out of my throat." He took another long drink.

"Thanks, sweetie," she acknowledged. "I know it's been frustrating for you this week."

"Aw, baby, that's not the half of it. There's just nothin' to haul. The plants are dead, the fields are turnin' to dust, and the wells are dryin' up. Some of these guys are havin' to slaughter their own herds of cattle 'cause they can't feed or water them. A lot of the younger farmers are talkin' about packin' it in, and some of 'em already have. They're already gone... poof. Said they just can't make it under these conditions and don't see anything promising in the forecasts. Banks can't carry 'em forever, and they have nothing but debt to fall back on.

"Even some of the bigger farms and ranches are starting to get shaky, but most of those old coots have other resources, and they'll just hold on for spite, if nothin' else."

He took another few sips from his glass, then continued.

"It takes a real gambler's mentality to have the guts to be a farmer any more. Mortgage everything but your first-born child to buy the equipment and all you need to get those seeds into the ground, then hope and pray that the weather cooperates so you can harvest a crop. Once it's harvested, you hope and pray that you can sell it for enough money to pay your bills and buy what you need so you can start the whole process over again. I swear, farmers are either masochists or they're stark-starin' crazy! I must be crazy, too,'cause I'm right out there with 'em waitin' to haul... What?... Dust?"

With that Bob finished his drink, then rose and poured himself another glassful of the sweet, citrusy mixture. There was more he needed to say, but in all his years with this wonderful woman, he had never intentionally hurt her, and he didn't want to now. Instead of returning to his chair, he sat down next to her on the couch. As he did so, Laura knew that whatever was coming was not going to be good, but she tried not to panic.

"Honey, I don't even know how to start this conversation. I know that I've really been thinkin' about this a lot while I've been on the road these last few weeks. Actually, I've been toying with this idea for quite a while now."

This REALLY doesn't sound good, her psyche almost screamed. *I can't imagine that he's been unfaithful. He's just too good a man even to consider such idiocy, but I have no clue where this conversation is headed.*

"Laura, you and I have spent our entire lives within fifty miles of this very house, with the exception, of course, when we lived in Lincoln and went to school." He smiled and cheered, "Go Big Red!"

Laura laughed. Lincoln was a wonderful college town. They went to school there when bleachers were still set aside in the football stadium end zone for the elementary school children in town. For fifty cents the little ones joined the *Knothole Club* and had their own special cheering and play section. They were corralled by fences and supervised by security, but the kids never even noticed, especially when the cheerleaders came over to lead them in cheers. If the cheerleaders wanted extra noise, they could always depend on the children to add to the enthusiasm, regardless of the score.

The University campus and the stadium were not far from downtown Lincoln where the hometown team was supported and celebrated in every store and restaurant around. The roar of the crowds, the play-by-play announcers, the band, the clamor of raucous celebration emanated from radios and speaker systems for those unfortunate souls who weren't able to make it to The Game. Of course, everyone wore red, *and* everyone knew the University's fight song, "There is... no place like.. Nebraaaska...!" Yes, that was a fun place to be.

"Laura, I think we need to start talkin' about the possibility of makin' a move ourselves..."

"You've got to be kidding, right?" she interrupted. Of all the possible subjects he could have brought up, that one was absolutely not even on the radar screen. She was stunned. "What in the world are you talking about?"

"Just hear me out for a bit. I think it's a topic we need to start considerin', not that it's somethin' we need to do next week. Honey, you and I are educated people, and we have been around this business all of our lives. This company managed to make it through the Great Depression, and we weren't doin' half bad holdin' on durin' this Great Recession. But if the forecasters and scientists are right about this atmospheric imbalance business, then what is comin' is a whole heck of a lot worse than all that other stuff put together."

He stopped and shrugged his shoulders in the darkening room. Then he leaned back and slouched down in the couch.

Laura said nothing although she was frightened. It had been quite some time since they had had a conversation of any depth or significance. It seemed that the garbage of life was always getting in the way of discussing those subjects that they thought really mattered. It didn't used to be that way. They would stay up all night having endless conversations about the mysteries of the universe and the meaning of existence and all those topics that youth seem to pontificate upon ad nauseum. But then the real world stepped in with bills and mortgages and dirty diapers...then school projects and little league...and...and...until most of the truly monumental discussions faded into memories.

It wasn't that Bob and Laura weren't close; actually, it was just the opposite. They were always discussing those subjects that needed to be addressed. It's just that important subjects were folded into the daily conversations of life while the two were preparing dinner or working in the yard together or driving to a football game. They always talked. Yet, everything about this conversation was different, and for the first time since she was a teenager, Laura thought she might really faint...the warm flush, the stars dancing across her eyes, the cotton in her ears so Bob's voice sounded muffled and far away, the darkening around her.

She had enough sense to put her wine glass down and take some deep breaths. The moment passed, hopefully without Bob noticing. *I need to pay careful attention to what he is saying,* she determined. *We obviously are headed into life-altering times.*

He could tell she was reeling from the discussion he had just opened, so he suggested that she turn around and place her head upon his lap. That way, he could see her face and stroke her hair and have his arm around her as he told her of his ideas. She thought that was an excellent suggestion, so she grabbed an afghan to cover her feet and lay down. Soon the blood was pumping to her brain again, and she began to feel better.

"The reason that we have to consider the possibility of a move are two-fold," he iterated. "For one thing, we cannot continue to support this business and all of its employees, who are, as you know, mostly our family members, on our savings and retirement. It's not good business sense, and it's just plain unhealthy for all of us from a personal perspective. You've seen the books. You know what I'm talkin' about."

Laura nodded, patted his arm, and agreed. "Yeah, I know you are exactly right. This drought and the low crop yields and the slow business did not all just happen overnight. We've been watching and feeling the impact of this for, what, three years now?"

"At least that," he responded. "But here's the second point, and, I'm afraid, the bigger issue. We don't know how long this thing is gonna last. My fear is that if we try to wait it out, and the forecasters and scientists are correct, then we run the risk of depleting our resources and making our chances for change much more difficult for everyone."

Bob and Laura were silent again. The large kitchen-family area at one point had been called a "country kitchen" by architects and designers and had been wonderfully suited for their raucous, always-hungry, always-on-the-go family. Right now though, it was dark and quiet. They could hear the *clu-clunk* of the ice maker in the freezer, soft voices outside saying "good night," car doors closing, engines starting, gravel under tires, and, finally, the squeak of the gate as it was closed and locked for the night.

With the rising of the moonlight came the melodious cacophony of night sounds, the hums, the buzzes, the croaks of the frogs, the crickets, the owl, the katy-dids — that other world that comes alive at night after the humans have gone inside. Neither of them rose to turn on a light; the rising moon began to transfuse the room with its own wistful, translucent illumination. Nothing more was needed.

After a few moments, Bob began to speak again, his voice was soft, thoughtful. He reminded her of a book they had read in school when they were young entitled *Night* by Elie Wiesel.

Early in the true story, Jewish Elie tries to convince his father that their family must flee Europe on the next boat for Haifa because of ongoing consequences to Jews at the hands of the Nazis. The father does not take the stories seriously and refuses to leave his home and his life's work behind for the sake of rumors. Of course, at the end, only Elie survives the concentration camps to tell the tale.

"I have often wondered," Bob pondered. "How do people know when to leave? Did you know that Albert Einstein left Europe and came to America in 1933 because he was already extremely uncomfortable with what Hitler and the Nazis were doing? Now granted, he was Einstein, after all. But, still...how could so many millions of people have been trapped by the insanity and killed, while so many others saw the same circumstances and said to themselves, 'It's time to go!'"

The memory of the book and the events surrounding it still brought tears to Laura's eyes. She, too, had wondered why so many people failed see the danger until there was no escape.

"Do you think we're missing the danger here?"

"I don't know, Laura, I don't know. I do know one thing. I trust C.B.'s judgment and insight. He is rational and a good, objective sounding board. I suggest we take a trip to North Carolina and have a conversation with C.B. and Vanessa. I'll tell him, in general, what we want to talk about, and I promise you, knowing C.B., he will have researched the topic thoroughly before we arrive. I can just see him handing us notebooks of information when we get there. Bar graphs. Charts. Ten-year studies of rainfall in Nebraska. Et cetera, et cetera, et cetera!"

With that, he chuckled, stretched, and lay with Laura in his arms. As the moonlight filtered through the sheers that billowed gently over the open windows, the breezes began to whisper away the heat of the day. The even, rhythmic *click-click* of the fan was a soporific, and soon, their soft, slow breathing was the only other sound that could be heard in the sleeping house.

§§§

Chapter Twenty-Two

Environmental Protection Agency, Washington, D. C.
Saturday, September 8, 2012, 2:03 PM

Alda Martin Starkey listened to the light, pleasant conversations of the small gathering in her office. Carter had invited his parents, and Alda always enjoyed their company, so ingrained with genteel southern hospitality. Patty and her husband had brought Kaitlyn, their oldest daughter, and one of her friends. She glanced over to see her husband's easy smile as the Admiral enjoyed a moment with Justin, Katy and several others.

Alda moved to the front, and the conversation quieted as she raised both hands and spoke to the group.

"As Deputy Director, I wish to congratulate Scott Wasson and Katherine Carson on the <u>State of the Heartland</u> dispatches. Their series of 'eyes on' reports will be made public this Monday, as will their association, and their subsequent disassociation, with the Agency. At that point, the responsibility for the release of the information will be mine alone, and that's the way I want it to be. This is a moment of celebration for the success of this summer's effort. The afternoon is ours, and tonight we will gather at that infamous Logan's Roadhouse where this adventure began three months ago."

Alda paused for a moment, then spoke again.

"Scott and Katy, thank you for taking these three months out of your lives. We can only be patient now. Perhaps your <u>State of the Heartland</u> series will arouse the American people. At the very least, we have made the effort to stir interest and concern regarding our imbalanced atmosphere.

"Now, I would ask both of our intrepid travelers to share their thoughts on their heartland experiences."

Scott Wasson stepped forward, and gestured toward Alda Martin Starkey.

"First of all, credit should be given where credit is due. In a few days, Katy will be in Indiana, diligently pursuing graduate studies, and I will be back in the U.K. Next summer, we both will have finished our academic endeavors, hopefully, and perhaps, start our life together. Meanwhile, this lady will undoubtedly be experiencing more than a little political heat. Please join me in applauding her staunch courage to, as she put it, 'step to the right side of history.' "

After a full round of heartfelt applause, he continued.

"Before we return to our afternoon's sociability, I feel obligated to comment on this 'right side of history.' Please join in at any time you feel moved to do so, and we'll spend as much time exchanging points of view as we wish.

"Let me begin by saying that it's obvious to me that science, with whatever evidence it has to support the facts of atmospheric imbalance and its causes, is unable to break through the political barriers created by certain specialized interests.

"Actually, the main point of contention has moved beyond the question of an increase in global temperatures. The argument put forth at the moment questions whether atmospheric imbalance is a result of human activity or not. The specialized interests contend that human activity is not responsible for the extreme weather events we now experience, but is attributable to cycles of change the planet has experienced many times in the past.

"They say, 'come on now, all the alarm is based on phony, doctored science, promoted by the scientific community, greedily eager to obtain government grant money.' So, evidently, there is no need to worry. We'll be fine continuing just as we are, burning ever more amounts of fossil fuels, producing ever more amounts of greenhouse gases. The political policy advocated by those beholden to the specialized interests? 'Let the E.P.A. go the way of the dinosaur' and 'Drill, baby, drill.'

"Now," Scott continued, "how do we get ourselves, and those around us, to the right side of history? First of all, how do we become aware of the truth of a matter? And then, how do we decide where we should stand? Most importantly, what is our responsibility to our fellow citizens when we have become aware of significant truth?

"In way of explanation, let me read this:

"'We were not born critical of existing society. There was a moment in our lives (or a month, or a year) when certain facts appeared before us, startled us, and then caused us to question beliefs that were strongly fixed in our consciousness — embedded there by years of family prejudices, orthodox schooling, imbibing of newspapers, radio, and television. This would seem to lead to a simple conclusion: that we all have an enormous responsibility to bring to the attention of others information they do not have, which has the potential of causing them to rethink long-held ideas. ' Howard Zinn, 2005.

"Alda Martin Starkey, you have become aware of the truth of atmospheric imbalance, and you have shouldered your responsibility to reveal that truth. Thank you for doing so. I'm proud, and I'm certain Katy is as well, to have been part of this summer's program."

Kaitlyn McCarty raised her hand, and Scott nodded toward her.

"We've started this school year in General Science with the topic of climate change. This year has had the most extreme weather in the January through September - spring, summer, and fall - since records were started in 1910. What we've been talking about is just what you said. Whether human activity is responsible for climate change or whether it is just a long-term cycle. What I wonder is, what if it is human activity that's the cause? Shouldn't we be doing something about what we do? Or, do we just take the risk that it's normal, and that it's all going to be O. K.?"

"Kaitlyn, let me answer your question this way," Scott responded softly. "Less than 1% of the world's scientists dispute atmospheric imbalance and global warming. Nor do they dispute that human activities are the major cause behind these changes. But, the latest polls show that only slightly over half, somewhere between 50 and 55%, of the American public are currently convinced of the validity of global warming and climate change.

"Three months ago at Logan's Roadhouse," he continued, "I urged Alda not to connect the dots, but to converge the dots, pile them on top of each other. Now everyone, all Americans, must do the same.

"Here's a dot: floods that covered thousands of acres last spring in the Mississippi River Valley, which may have been several states away from you. Another dot: wildfires, uncontrolled for weeks at a time, burning hundreds of thousands of acres this summer, happened mostly out in the western states. And other dots: severe drought that has parched the states of Nebraska, Kansas, Oklahoma, Texas, and EF-5 tornadoes in the Midwest and southeastern states. All these events may have happened hundreds, perhaps, thousands of miles from your home."

Kaitlyn was a bit confused, but she continued to listen attentively to Scott's explanation.

"In reality, these are not unconnected dots, separated by weeks, or months, or a season. In terms of time, that's the blink of the eye. They are not unconnected dots in terms of distance from you, either. One state away from you, or a thousand miles, in terms of distance, it doesn't matter. It's all space on the surface area of our planet, our Earth.

Eliminate the time factor, eliminate the distance factor. Converge the extreme weather event dots. That's our climate, and it's not normal now, because our natural systems are increasingly out of balance.

"At Logan's Roadhouse, I used an example of a kettle with water at room temperature being heated for tea, that when the heat conditions increase enough, the water begins to simmer before the boil. That kettle is our planet, we are in the kettle, and it's beginning to simmer.

"The people who monitor and record weather activity have known all this for years." Scott tried to explain. "They are aware that while we have had only one degree of temperature rise in the lower forty-eight states, there has been a five-degree increase in Alaska, and eight degrees in the polar region. They know that very soon there will be no ice sheet at the North Pole, that the insidious danger, generally unrecognized, unacknowledged, is sea level rise, happening much faster than expected.

"But now, it's not the scientists with jagged lines slanting up on graph paper that present compelling evidence. It's no longer 'I don't know what to think, one way or the other' for the people in the heartland of America. Now it's extreme weather, and it's in their faces. And pardon me, Mrs. Starkey, but it is no longer a time to be patient."

Scott's frustration and impatience had become evident. He was no longer speaking just to Kaitlyn, and everybody was listening with full attention.

"The problem is getting worse. China is bringing a coal-fired power plant on line at the rate of roughly one a week. Efforts to control, much less reduce, this problem are completely ineffectual. So, we're continually playing Russian roulette with the planet. With every spin, we up the risk of more frequent and intense natural catastrophes.

"The multi-national corporations, responsible only to maximize profits at any cost, have been enormously successful at continuing spin after spin after spin. They are masters of disinformation, and have succeeded brilliantly in sowing and maintaining confusion in the minds of everyday Americans. To put it bluntly, the fossil fuel industries, particularly, are ruthless parasites on the body of mankind. They should keep in mind that when enough life is sucked from the host, the host dies, and the parasites perish with it. They are treading a fine line.

"At the end of the day, the negative impact their greed has inflicted on our world is a massive problem that must be dealt with as quickly as possible. It is an issue that transcends political lines, political partisanship, political gridlock. We, the people, will determine our own destiny, either by our action, or by our inaction. Therefore, we must persuade our elected representatives to address the problem, to bend their will to the needs of all humanity, not just to the moneyed interests of multi-national corporations. We must make their political survival more dependent on our petitions, recalls, referendums, and ballots than those other interests.

"What we need, must have, and soon, is the elimination of the gaseous chemical imbalance already present in our atmosphere. We don't sense that problem in our every day lives. But it's there, and manifests itself as increasingly extreme weather events. For the sake of all on earth, America must lead, must mandate a national goal, energize a national effort, and establish a national program empowered to return and maintain the natural balance of our planet. Simply stated, the responsibility is on our shoulders, and there is no time to lose!"

Alda was taken aback as she listened to the passionate statement of this brilliant young scientist. *His message is visionary, but his rhetoric is too inflammatory,* she reflected silently. *He's not afraid to be bold, but I have a role to play here if that message is to be effectively received by the public.* Mrs. Starkey's husband saw his wife silently sit up straighter in her chair and lift her chin. Without knowing what it was, the Admiral knew that Alda Martin Starkey had just come to a powerful decision. As they both listened to Scott detail what needed to be done, he also knew he didn't need to ask, that her resolution would be revealed with time.

§§§

Chapter Twenty-Three

North Carolina 2012
"The Fire Pit"

There is nothing like the phrase "company's coming" to get a house sparkling and in tip-top shape. The length of the visit determines the extent to which this adage applies, and C.B. and Vanessa's house was no exception. Some people are immaculate housekeepers and never need hours of warning before even the slightest of drop-ins, but that was not Vanessa. She and C.B. always had projects going that required emptying out closets or cabinets or sometimes whole rooms. They enjoyed painting and decorating and finding better ways to use the furniture they had both inherited and purchased along the way.

Then there were the repairs and replacements that had to be completed, such as replacing the old, barely functional doorknobs in the bedrooms with the new crystal ones Vanessa had gone to such lengths to special order. Somehow those doorknobs had never made it to the top of C.B.'s *To Do* list. No, it was a matter of priority: which crisis was the most urgent?

So when Bob and Laura notified them of an important week-long visit they wished to make, the intrepid Carolinians were thrilled.

Then they sat down with pen and paper to map out which projects absolutely had to be accomplished and which items could be moved to the "if time permits" column. Even the man of the house had to admit that working doorknobs was a priority for guests, much to Vanessa's relief.

C.B. had recognized the undertones of concern and anxiety in his friend's voice when Bob called and asked if C.B. and Vanessa had time for a visit. They both knew that their conversations would be far more substantive than perhaps even Laura realized. Vanessa, too, surmised that they weren't traveling halfway across the country just to roast marshmallows in the fire pit. There weren't even any good karaoke places around for Bob to show off his considerable talent.

No, she concluded. *Whatever is going on, it's my job to help both of them relax and figure out what they need to figure out. That means that I have to finish my projects in time to be relaxed when they get here. Besides, I can't wait to see them!*

Mid-October was a wonderful time of the year for Bob and Laura to arrive in Statesville. Unlike Nebraska, rainfall in the North Carolina foothills had been plentiful, and many of the trees were just changing into their vivid fall colors. The days were still warming into the low 70's, but the chilly nights were reaching temperatures as low as the 30's and 40's. It was great fire pit weather, and they were all anxious to build a fire and have a cook-out.

The first two days were filled with light, catch-up conversations about life and family, looking at pictures, and preparing meals. The guys worked on the grounds. They cut up a tree that had fallen and carted extra wood and leaves down to the glen, where the pit was located.   Then they took care of whatever last-minute details were necessary for the cook-out and went inside. To their delight, they were greeted with hot chili topped with cheese and onions, cornbread, and their choice of beverages. It was a pleasant way to end a pleasant day. They all had needed the time out.

By the third day, the four felt like a team and were ready to tackle the business that had brought the Millers from Nebraska to the Carsons' doorstep in North Carolina. Early morning rain and clouds had threatened, but by noon, the skies had cleared leaving sunshine and a soft breeze. The men started the fire and set up the serving table in the glen.

The women collected the food and everything else they could think of, loaded it all in Vanessa's "catering wagon," and wound their way down toward the fire pit, laughing because they kept thinking of more stuff they might need.

"It's like packing," one of the guys sputtered as he howled with laughter when he saw them lumbering down the path. "She never knows what to take, so she takes it all!" "Absolutely!" the other one, also hooting at the ladies, agreed. As they settled down and settled in with drinks and plates of food, the talking began.

"It's curious," Bob reflected as he finished a bit of potato salad. "I think I know how a general must feel as he is preparing to deploy his troops. Where to place them? When to send them? How to get them there? How to support them once they arrive? And during the entire decision-making process... knowing that a miscalculation could be detrimental to everyone... an error in judgment could cost lives. Responsibility can be such a heavy burden."

No one spoke for a few minutes. The trees sang with the voices of cardinals, wrens, black-capped chickadees, goldfinches. The red-shouldered hawk could occasionally be seen circling overhead when it wasn't perched stealthily on a low-lying branch down the path. Even the creek, while not full, had enough water in it to be heard in the quiet of the late afternoon.

Finally, C.B. rose and began to stoke the fire with the six-foot tool he had crafted from a large limb the previous winter. The flames rose to his commands as he deftly poked and prodded the smoldering mass. He put more wood on the flames and began to speak to Bob.

"You know, Bob," C.B. seemed to be speaking to the fire more than to Bob at that moment. "Not every leader goes to a military academy, then has stars pinned to his chest and is told, 'Now, you are a leader. Now, you are capable of making wise leadership decisions.' Wouldn't it be wonderful if that were always the case?"

A chorus of, "Boy, isn't that the truth." and, "Don't we wish!" and "If only..." was heard around the pit.

"But, Bob," C.B. continued, "there are also those who are thrust into leadership positions without ever seeking or even wanting to be in front. They are the ones whose courage, history and integrity seem to provide that intangible quality necessary so they may positively lead the way for others."

Bob rose and began to help put more wood on the fire. Laura and Vanessa were glad they had on jeans and sweaters as the temperature was beginning to drop rapidly in the twilight.

"You are so right about being thrust to the front without wanting to be," Bob laughed as he responded. Until recently, Bob had been surprisingly reserved, even though he had been called upon to make tough business decisions along the way. Finding out that people enjoyed his singing voice and then performing before crowds had given him a self-confidence he had not known before.

C.B. laughed at him. "Bob, you stand up and sing in front of large groups of people at a time without batting an eyelash, while the rest of us would absolutely pass out with fear and trepidation. Vanessa has had a beautiful voice and has been chosen for select choirs and performing groups all of her life. Yet she has never once had the courage to perform a solo. It just takes a certain kind of confidence to put yourself on the line like that."

"Aw, she just hasn't had the experience, that's all," Bob responded as he took a sip of cold water. "In time and with the same encouragement that I have received, she could do it, too."

"Actually, my dear friend," C.B. said as he turned and looked at his visitor. "You have just made my point."

Both Bob and Laura looked at him quizzically.

"Look at it this way. You have been leading your tribe now ever since you took over the company. That was... what?... thirty-five years ago? Something like that? Anyway, in that time you, and Laura, I might add, have earned the respect and admiration of the members of your extended family, as well as the other workers in the company. I've seen it every time I come for a visit.

"That didn't happen overnight, but over years of honesty and fairness and good decision-making and wise choices. They know and trust that whatever choices and decisions you make will be thoughtful and well considered."

Bob sighed and smiled, "Thanks, friend. Deep down, I think I've known that all along. But, what I am considering asking them all to do frightens even me. What if I'm wrong? Here, let me lay it out for you."

The two shared a friendship that went back to early elementary days and childhood sandboxes. As they sank deeper into their conversation, Vanessa gently tapped Laura on the arm and nodded to the food sitting on the table. It was never wise to leave open containers of food sitting out in the woods after dark, so, together, they silently filled the catering wagon. Vanessa put more wood on the fire, then they followed the path up the hill to the main house. They hastily put away what had to be put away, then grabbed jackets and hurried back down the hill.

Their men were still deep in discussion, but were glad to see them return and stopped long enough to refill wine glasses or hot cocoa mugs. They again stoked and prodded the fire, then drew Laura and Vanessa into their conversation.

C.B. and Bob were both anxious to catch their wives up on what had transpired thus far. They had been talking about the imbalanced atmosphere and how dramatically it was impacting their lives, especially Bob and Laura's.

They had talked, too, about Nick and Kara. Nick had been gathering information for quite some time now from his scientist colleagues who were all saying basically the same thing. The new, stronger, more violent weather patterns were not anomalies, but rather, he and his colleagues had concluded, given all the data they had acquired, the new norms.

C.B. also told Bob that, even with all of his scientific data, Nick and Kara were still in a quandary about their own circumstances. Having all the statistics pointing toward a certain decision didn't help in making that decision any easier, especially when it constituted leaving behind everything you had ever known - your home, your family, your friends.

One of the reasons Bob had decided to make the trip was because he felt certain that Nick's scientific bent and professional knowledge would be made available to them and might assist them in seeing the bigger picture. He was somewhat surprised to discover that Nick and Kara were struggling with the same decisions that he and Laura were.

C.B. went on to explain that Kara's parents were not the only ones resisting the veracity of the scientific information continually being released on the deteriorating health of the planet. Nick's father and stepmother, too, were skeptics of the first magnitude. "Scientists are just trying to scare the public so they can get grant money for their pet projects." or "Scientists are just trying to discredit this candidate or that candidate because they don't like his or her policies." or "The pictures of the melting ice caps aren't real." They even believed Nick's own figures were made up. They knew all the arguments against believing the truth of what was occurring before their very eyes.

Vanessa finally spoke up. She didn't want the conversation to be reduced to a political discussion. The stakes were too high and the decisions too important for their dear friends and for the precious families that she and C.B. were also worried about.

"I'm not sure what conclusions you all have come to," she began, "but, let me share with you what Nick and I have been talking about."

Everyone else quieted, and Vanessa began again in earnest. "I think we all agree that we still have some time before decisions absolutely *must* be made. Right?"

Affirmative responses were heard all around.

"Good, OK. Now some scientists are saying that we may still have decades before we can no longer do anything about the atmospheric imbalance, while others are predicting that the time is very short - maybe only a few years.

"Personally, I want to be as prepared as I can be so I can survive as long as possible, whatever happens. I just cannot believe that someone or some group isn't capable of coming up with a method, a means, a machine, some way to remove the garbage that is in the atmosphere before humanity ceases to exist. If we could send men to the moon using computers with less power than our cell phones, surely we can figure this one out."

No one spoke. They all knew she had more to say.

"Here are some things that Nick and I came up with. You've probably come up with the same things, but here goes:

1) Wait out the winter to see what happens with the drought. If the rain and snowfall are sufficient, then put any contingency plans that you can draw up this winter on the shelf.

2) Consider your resources. The more you tap into your savings and your retirement, the less you will have to relocate and re-establish a business, or, Heaven forbid, go into survival mode.

3) By leaving the Homestead while you still have the wherewithal to continue paying the taxes on the property, it will be there when you return after the drought is over or when the crisis has been resolved. If you wait too long without having any work, you run the risk of losing the land and your livelihood, because you won't have the means to make a move.

4) If you wait too long and leave at the time when so many others in the heartland have also decided to leave, you will become a part of what is known as a *climate migration*. It has already occurred in a number of places around the globe. Hundreds of thousands of people on the

move to get away from drought, floods, heat, and always, starvation. They have nowhere to go *to*, only *from*.

5) If you leave before the *migration,* there will still be food and housing and business opportunities available in another part of the country, but don't wait too long."

With Vanessa's rather grim pronouncements, no one said anything for some moments.

Then Laura spoke, "As I listened to you, I kept asking myself over and over, how do we know whom to believe, and how do we know when to go?"

C.B. ventured into the conversation with his own suggestions. "First of all, start doing your own research. You don't have to be a scientist to learn how to read the Google maps of ocean temperatures and currents. I feel certain you both know how to read weather maps — high pressure systems, low pressure systems, and so on?"

"Absolutely," Bob responded. "We've gotta know what's comin'."

"Yeh, so all you need to add to that is a notebook with graph paper. Start keeping track on a daily basis, you know, highs, lows, rainfall amounts, winds, *and* leave room to record any anomalies, unusual weather events, that occur that day — maybe just your area, your region, or maybe the whole

U.S. Then start making your own comparisons from past years using your own state reports. That's pretty basic, and you won't have all the fancy equipment. But, you will be able to choose your own sources from everything on the web. I would even encourage you to go to scientific sources outside the U.S.

"But, whatever you do, my friends," he finally said, "have your Plan B ready to go if you need it."

By this time the fire had burned down to embers, and the critter who normally used the glen for his nighttime escapades was chattering furiously at them. It was, indeed, time to turn the forest back over to the creatures of the night.

§§§ *Fini* §§§

About the Authors

The Kettle Begins to Simmer 2012
The Imbalanced Atmosphere Chronicles
A Connective Collaboration of Fiction

Deborah Lapping Dorsey first became interested in the weather when, as a child, she would look for Pegasus flying among the clouds. Her love of the language and literature led her to the classroom where she taught high school English for thirty years. While in Texas, she was selected as one of the top three finalists for the 1999 Texas Secondary Teacher of the Year. One of her Teacher of the Year essays later became the basis for the "No Child Left Behind" concept.

Mrs. Dorsey received a Bachelor of Arts Degree in English and a Master of Science in Education/Reading Specialist Degree from Pittsburg State University in Kansas. She is retired and lives in North Carolina with her husband William. She enjoys trying to identify the myriad birds in their woods. She also treasures making her eighties-something mother laugh.

William Dorsey became particularly interested in the impact of the changing atmospheric conditions as it relates to weather when, in 2005, he was forced to drive back roads into New Orleans to rescue his daughter from hurricane Katrina's onslaught. Since that time, he has actively supported awareness of the environmental crisis that has resulted from the planet's imbalanced eco-systems.

After four years of active duty in the United States Marine Corps in the 1960s, Mr. Dorsey received both a Bachelor of Science and a Master of Science in Industrial Technology from Pittsburg State University in Kansas. Now retired, he enjoys writing and the outdoor experiences in his three-and-a-half acres of woodlands.

www.ingramcontent.com/pod-product-compliance
Lightning Source LLC
Chambersburg PA
CBHW030010290326
41934CB00005B/281